FRACTALS AND CHAOS
AN ILLUSTRATED COURSE

Related Titles

Universality in Chaos, 2nd edition
P Cvitanović (ed)

Introduction to Chaos and Coherence
J Frøyland

Hamiltonian Dynamical Systems
R S Mackay and J D Meiss (eds)

FRACTALS AND CHAOS

AN ILLUSTRATED COURSE

Paul S Addison

Napier University, Edinburgh

Institute of Physics Publishing
Bristol and Philadelphia

British Library Cataloguing-in-Publication Data

A catalogue record for this book is available from the British Library

ISBN 0 7503 0399 9 (hbk)
     0 7503 0400 6 (pbk)

Library of Congress Cataloging-in-Publication Data

Addison, Paul S.
      Fractals and chaos : an illustrated course / Paul S. Addison.
          p.   cm.
      Includes bibliographical references and index.
      ISBN 0–7503–0399–9 (alk. paper). –– ISBN 0–7503–0400–6 (pbk. : alk. paper)
      1. Fractals. 2. Chaotic behavior in systems. I. Title.
      QA614.86.A23 1997
      514'.742--dc21   97–18158
                                                                                     CIP

Reprinted with minor corrections 2001

This printing is a digital reproduction of the original book and this may result in a slight loss of clarity in some of the figures.

Published by Institute of Physics Publishing, wholly owned by The Institute of Physics, London

Institute of Physics Publishing, Dirac House, Temple Back, Bristol BS1 6BE, UK

US Office: Institute of Physics Publishing, The Public Ledger Building, Suite 1035, 150 South Independence Mall West, Philadelphia, PA 19106, USA

Typeset in TEX using the IOP Bookmaker Macros
First printed in the UK by Bookcraft Ltd, Bath
Reprinted in the UK by Copy-Book Publishing, Gloucester

I dedicate this book to my four best friends:

Stephanie Addison,   *Josephine Addison,*
Stanley Addison,   *Michael Addison;*

wife, mother, father, son.

Contents

Preface

The aim of this textbook is to provide the reader with an elementary introduction to fractal geometry and chaotic dynamics. These are subjects which have attracted immense interest throughout the whole range of numerate disciplines, including science, engineering, medicine, economics, and social science, to name but a few. The book may be used in part or as a whole to form an introductory course in either or both of the subject areas. The text is very much 'figure driven' as I believe that illustrations are extremely effective in conveying the concepts required for comprehension of the subject matter of the book. In addition, undue mathematical rigour is often avoided within the text in order to provide a concise treatment of specific concepts and speed the reader through the subject areas.

To allow the reader a steady progression through the book, without too much jumping about from chapter to chapter, I have attempted to order the topics within the text in as logical a sequence as possible. Chapter 1 provides a brief overview of both subject areas. The rest of the book is split into two parts: the first (chapters 2–4) deals with fractal geometry and its applications, while the second (chapters 5–7) tackles chaotic dynamics. Many of the methods of fractal geometry developed in the first half of the book will be used in the characterization (and comprehension) of the chaotic dynamical systems encountered in the second half of the book. Chapter 2 covers regular fractals while chapter 3 covers random fractals. Chapter 4 goes off at a slight tangent to investigate the fractal properties, and usefulness, of fractional Brownian motions (fBms). Initially conceived as one or two sections within chapter 3's coverage of random fractals, it soon became obvious that such an important topic deserved its own chapter. I believe that fBms have a lot to offer the scientific community, not least in the modelling of non-Fickian diffusion and natural surface roughnesses. The absence of fBms, in any detail useful to the scientist or engineer, is conspicuous in many texts. However, the reader wanting to move quickly through the text from fractals to chaos may skip chapter 4. Chapter 5 deals with chaos in discrete dynamical systems. Chapter 6 covers chaos in continuous dynamical systems, and the tools necessary for the characterization of chaos are detailed in chapter 7. Among other things, chapter 7 links the fractal geometry of chapters 2 and 3 with the chaotic dynamics dealt with in chapters 5 and 6. In appendix A, a computer code for demonstrating chaos in the Lorenz equations is provided for use in the questions at the end of chapters 6 and 7. Appendix B illustrates the application of some of the techniques learned in chapters 6 and 7 to real experimental systems in which chaos has been observed. Seven systems are detailed briefly in this appendix and these are selected from a broad range of scientific and engineering areas.

The text is written at an elementary undergraduate level. It is based upon a university course, designed and run by myself, which may be taken by any student from a numerical discipline. The multidisciplinary nature of the topics within the book make it ideal for the basis of an elective course, with a university-wide appeal. In addition to undergraduate students, the book should also be of benefit to postgraduate students, researchers and practitioners in numerate disciplines who have no previous knowledge of the topics and who want a quick introductory overview at a reasonable mathematical level. An attempt has been made to relate the text to the reader's own area of interest, and to that end, a substantial further reading section is provided near the end of each chapter. These further reading sections should point the reader to more advanced sources of information specific to his or her own field of interest. Although I have included references from as many subject areas as possible, the literature is vast and diverse and there will inevitably be many omissions. I have, however, attempted to provide the more important historical references together with some of the more recent ones.

After the further reading section in each chapter there are revision questions and further tasks for the reader to attempt. The solutions are provided at the end of the book. Within the text itself, I have attempted to be as consistent with the nomenclature as possible; however, often this is not possible (especially in chapter 7) as many symbols have been used in the literature for historical reasons, and I have kept these in their traditional form. In addition, to assist the reader I have used *italics* for mathematical text, key phrases and quotations; **bold** for keywords; and <u>underlining</u> for emphasis (used sparingly).

I would like to take this opportunity to thank the many people who helped me in the completion of this book. I would like to thank David Low, David Tritton and Iain Lawson for taking the time to read over the draft manuscript and providing many helpful suggestions. I would like to thank Andrew Chan for his comments on the draft manuscript; for his Runge–Kutta routine used in the program in appendix A; and, not least, for his great help as a PhD supervisor when I first became involved in fractals and chaos many moons ago (or so it seems). Thanks are also due to Ron Hunter for his generous assistance in the drawing of some of the figures in chapters 2 and 3, and for his technical assistance which allowed me to draw many of the other figures in the book. In the same vein, thanks also to Peter McNeill and Donald Ross. Thank you to Jamie Watson who provided the algorithm for the generation of the DLA figure in chapter 4.

Special thanks to both my wife Stephanie for her support and encouragement during the writing of the text and my son Michael who, although he cannot speak as yet, encouraged me greatly with his gummy smiles (and latterly his sparsely toothed smiles). Special thanks also to my parents for their unending help and support during the writing of the text.

Paul S Addison
March 1997
p.addison@napier.ac.uk

Chapter 1

Introduction

1.1 Introduction

The twin subjects of fractal geometry and chaotic dynamics have been behind an enormous change in the way scientists and engineers perceive, and subsequently model, the world in which we live. Chemists, biologists, physicists, physiologists, geologists, economists, and engineers (mechanical, electrical, chemical, civil, aeronautical etc) have all used methods developed in both fractal geometry and chaotic dynamics to explain a multitude of diverse physical phenomena: from trees to turbulence, cities to cracks, music to moon craters, measles epidemics, and much more. Many of the ideas within fractal geometry and chaotic dynamics have been in existence for a long time, however, it took the arrival of the computer, with its capacity to accurately and quickly carry out large repetitive calculations, to provide the tool necessary for the in-depth exploration of these subject areas. In recent years, the explosion of interest in fractals and chaos has essentially ridden on the back of advances in computer development.

The objective of this book is to provide an elementary introduction to both fractal geometry and chaotic dynamics. The book is split into approximately two halves: the first—chapters 2–4—deals with fractal geometry and its applications, while the second—chapters 5–7—deals with chaotic dynamics. Many of the methods developed in the first half of the book, where we cover fractal geometry, will be used in the characterization (and comprehension) of the chaotic dynamical systems encountered in the second half of the book. In the rest of this chapter brief introductions to fractal geometry and chaotic dynamics are given, providing an insight to the topics covered in subsequent chapters of the book.

1.2 A matter of fractals

In recent years, the science of fractal geometry has grown into a vast area of knowledge, with almost all branches of science and engineering gaining from the new insights it has provided. Fractal geometry is concerned with the properties of fractal objects, usually simply known as **fractals**. Fractals may be found in nature or generated using a mathematical recipe. The word 'fractal' was coined by Benoit Mandelbrot, sometimes referred to as the father of fractal geometry. Mandelbrot realized that it is very often impossible to describe nature using only Euclidean geometry, that is in terms of straight lines, circles, cubes, and such like. He proposed that fractals and fractal geometry could be used to describe real objects, such as trees, lightning, river meanders and coastlines, to name but a few.

There are many definitions of a fractal. Possibly the simplest way to define a fractal is as an object *which appears self-similar under varying degrees of magnification. In effect, possessing symmetry across scale, with each small part of the object replicating the structure of the whole.* This is perhaps the loosest of definitions, however, it captures the essential, defining characteristic, that of **self-similarity**. A diagram is possibly the best way to illustrate what is meant by a fractal object. Figure 1.1 contains sketches of two naturally occurring 'objects': an island coastline and a person. As we zoom into the coastline, we find that its ruggedness is repeated on finer and finer scales, and under rescaling looks essentially the same: the coastline is a fractal curve. The person, however, is not a self-similar object. As we zoom into various parts of the body, we see quite different forms. The hand does not resemble the whole body, the fingernail does not look like the hand and so on. Even viewing different parts of the body at the same scale, say the hand and the head, we would see that again they are not similar in form. We conclude that a person is not a fractal object. It is interesting to note at this stage that, although the body as a whole is not a fractal object, recent studies have attempted, with some success, to characterize certain parts of the body using fractal geometry, for example, the branching structure of the lung and the fine structure of the neuron (brain cell).

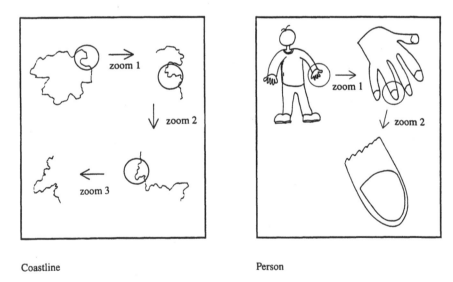

Coastline   Person

Figure 1.1. Fractal and non-fractal objects.

Figure 1.2 contains four **natural fractals**: the boundary of clouds, wall cracks, a hillside silhouette and a fern. All four possess self-similarity. The first three natural fractals possess the same statistical properties (i.e. the same degree of ruggedness) as we zoom in. They possess **statistical self-similarity**. On the other hand, the fern possesses

Cloud outlines   Wall cracks

Hillside skyline   Fern tip

Figure 1.2. Natural fractal objects.

exact self-similarity. Each frond of the fern is a mini-copy of the whole fern, and each frond branch is similar to the whole frond, and so on. In addition, as we move towards the top of the fern we see a smaller and smaller copy of the whole fern. The fractals of figure 1.2 require a two-dimensional (2D) plane to 'live in', that is all the points on them can be specified using only two co-ordinates. Put more formally, they have a Euclidean dimension of two. However, many natural fractals need a 3D world in which to exist. Take, for example, a tree whose branches weave through three dimensions; see the tree branching in 3D in figure 1.3 (if you can!). Fractals themselves have their own dimension, known as the **fractal dimension**, which is usually (but not always) a non-integer dimension that is greater than their topological dimension, D_T, and less than their Euclidean dimension D_E (see chapter 2). There are many definitions of fractal dimension and we shall encounter a number of them as we proceed through the text, including: the similarity dimension, D_S; the divider dimension, D_D; the Hausdorff dimension, D_H; the box counting dimension, D_B; the correlation dimension, D_C; the information dimension, D_I; the pointwise dimension D_P; the averaged pointwise dimension, D_A; and the Lyapunov dimension D_L. The last seven dimensions listed are particularly useful in characterizing the fractal structure of strange attractors associated with chaotic dynamics.

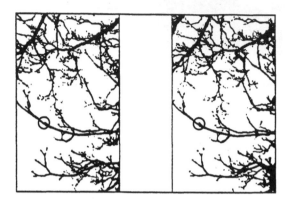

Figure 1.3. Tree branching in 3D. To see the 3D image, illuminate the page with a good even source of light: daylight is by far the best. Keeping the page still, view the images from a distance of 15–20 cm, let your eyes relax and try to merge the two circles. After merging, the image should come into focus within a few seconds. Once focused, let your eyes wander around the image to see 3D. The technique needs a little practice: the trick is to focus on the merged image without the two constituent images diverging, but persistence usually pays off.

We make one more important distinction between fractals which are self-similar everywhere and those which are self-similar only if we look in the right place. Examples are given in figure 1.4. The figure contains three **mathematical fractals**, these are: a logarithmic spiral, a binary tree, and a Sierpinski gasket. We see self-similarity in the logarithmic spiral of figure 1.4(*a*) only if we zoom into its point of convergence. The part of the spiral contained within box A contains the point of convergence, hence infinitely many scaled copies of the spiral exist within this area. However, the part of the spiral within box B does not contain the point of convergence and hence does not contain scaled down replicas of the whole log spiral. The binary tree (figure 1.4(*b*)) is simple to construct mathematically: we simply add further, scaled down, T-shaped branches to the ends of the previous branches. After an infinite number of branch additions we have the binary tree. As we zoom into the branches of the binary tree we see more and more detail, consisting of exactly self-similar copies of the whole tree. Hence, it is a fractal. However, the self-similarity of the binary tree (figure 1.4(*b*)) is only evident if we zoom into one of its branch ends. The circled area A contains one such branch end, which is an exact copy of the whole tree scaled down by one eighth. Contrast this with the part of the tree contained within the circled area B which is not a scaled down copy of the whole tree. The Sierpinski gasket of figure 1.4(*c*) (the construction of which is detailed in chapter 2) is self-similar everywhere. No matter where we zoom into the gasket, we will see further copies of the whole gasket. This property is known as **strict self-similarity** and the Sierpinski gasket is a strictly self-similar fractal. In this book we will concentrate on strictly self-similar fractals. In figure 1.2 the cloud boundary, wall

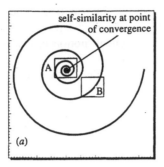

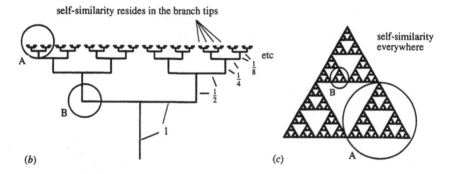

Figure 1.4. The nature of self-similarity. (*a*) The log spiral is self-similar only at its point of convergence. (*b*) The binary tree is self-similar only at the branch tips. (*c*) The Sierpinski gasket is self-similar everywhere.

crack, and hillside skyline are strictly self-similar, whereas the fern of figure 1.2 and the tree of figure 1.3 are only self-similar at their branch ends.

One last point worth noting is that even the best examples of natural fractals do not possess self-similarity at all scales, but rather over a sufficiently large range to allow fractal geometric methods to be successfully employed in their description. On the other hand, mathematical fractals can be specified to infinite precision and are thus self-similar at all scales. The distinction between the two is usually blurred in the literature, however, it is one worth remembering if you intend using, in a practical situation, some of the methods from fractal geometry learned from this text.

1.3 Deterministic chaos

Oscillations are to be found everywhere in science and nature. The mechanical engineer may be concerned with the regular oscillation of an out of balance drive shaft; the civil engineer with the potentially disastrous structural vibrations induced by vortex shedding on a bridge deck; the electrical engineer with the oscillatory output from nonlinear circuits; the chemist/chemical engineer with the regular cycling of a chemical reaction; the geologist/geophysicist with earthquake tremors; the biologist with the cycles of growth and decay in animal populations; the cardiovascular surgeon with the regular (and more so, irregular) beating of the human heart; the economist with the boom–bust cycles of the stock market; the physicist with the oscillatory motion of a driven pendulum; the astronomer with the cyclical motion of celestial bodies; and so on. (The list is extensive and diverse!)

Dynamical oscillators may be classified into two main categories: linear and nonlinear. In general, all real systems are nonlinear, however, very often it is the case that, as a first approximation to the dynamics of a particular system, a linear model may be used. Linear models are preferable from a scientist's point of view as typically they are much more amenable to mathematical analysis. (Hence the disproportionate number of linear systems studied in science.) Nonlinear systems, in contrast, are much more difficult to analyse mathematically, and, apart from a few exceptions, analytical solutions are not possible for the nonlinear differential equations used to describe their temporal evolution. In addition, only nonlinear systems are capable of a most fascinating behaviour known as **chaotic motion**, or simply **chaos**, whereby even simple nonlinear systems can, under certain operating conditions, behave in a seemingly unpredictable manner.

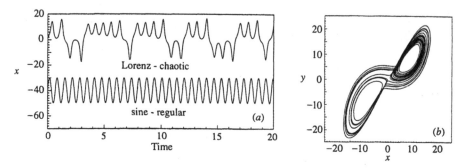

Figure 1.5. Chaos and regularity. (*a*) Time series of the chaotic Lorenz model and a periodic sinusoidal waveform. (*b*) Phase portrait of the Lorenz strange attractor.

In 1963 Edward Lorenz published his work entitled 'Deterministic nonperiodic flow' which detailed the behaviour of a simplified mathematical model representing the workings of the atmosphere. Lorenz showed how a relatively simple, **deterministic** mathematical model (that is, one with no randomness associated with it) could produce apparently unpredictable behaviour, later named chaos. The Lorenz model (see chapter 6) contains three variables: x, y and z. Figure 1.5(*a*) shows a chaotic time series output of the x variable of the Lorenz model. Notice that there is a recognizable structure to the time series: first the system oscillates in the positive-x region for a couple of oscillations, then it switches over to the negative x-region for a couple of oscillations, then back to the positive x-region for a few oscillations, and so on. However, the system never exactly repeats its behaviour. It would not matter how long we let the Lorenz model run for, we would never come across a repetition in the waveform. It is this aperiodic behaviour that is known as chaos. Compare it to the regular, periodic oscillations of the sinusoidal waveform (plotted below the Lorenz output) which repeats itself exactly and indefinitely. By plotting the Lorenz variables against each other, rather than against time, we can produce compact pictures of the system's dynamics. In two dimensions these are known as phase portraits. The phase portrait for the Lorenz time series is shown in figure 1.5(*b*). Starting the Lorenz system from many initial conditions produces phase portraits all of the same form: the system is attracted towards this type of final solution. Figure 1.5(*b*) is then a plot of the long term behaviour of the Lorenz system and is known as the attractor of the system. If we zoom into the fine scale structure of the attractor for the chaotic Lorenz system we see that it has a fractal structure. The attractors for chaotic systems which have a fractal structure are termed **strange attractors**. The fractal structure of strange attractors may be examined using one or more of the definitions of fractal dimension mentioned in the above section. (See chapter 7 for more details.)

Chaos has now been found in all manner of dynamical systems; both mathematical models and, perhaps more importantly, natural systems. Chaotic motion has been observed in all of the 'real' oscillatory systems cited at the beginning of this section. In addition, many common qualitative and quantitative features can be discerned in the chaotic motion of these systems. This ubiquitous nature of chaos is often referred to as the **universality** of chaos.

1.4   Chapter summary and further reading

1.4.1   Chapter keywords and key phrases

fractals	*self-similarity*	*natural fractals*
statistical self-similarity	*exact self-similarity*	*fractal dimension*
mathematical fractals	*strict self-similarity*	*chaotic motion/chaos*
deterministic models	*strange attractors*	*universality*

1.4.2   Further reading

Non-mathematical treatments of the history and role of fractals and chaos in science, engineering and mathematics can be found in the books by Gleick (1987), Stewart (1989), Briggs and Peat (1989), Lorenz (1993) and Ruelle (1993). Also worth consulting is the highly readable collection of non-mathematical papers by leading experts from various fields edited by Hall (1992). The explosion in the number of scientific articles relating to chaos and fractals is shown graphically by Pickover (1992). Simple computer programs to generate a range of fractal and chaotic phenomena are given in the text by Bessant (1993). Other forms of medium worth consulting are the videos by Barlow and Gowan (1988) and Peitgen *et al* (1990), and also the freeware software package FRACTINT, widely available on the internet, user friendly, and of excellent quality. In fact, a search on the world wide web using the keywords 'fractal' or 'chaos' should yield a large amount of material, much of which is at a reasonably elementary level.

Chapter 2

Regular fractals and self-similarity

2.1 Introduction

In this chapter we will examine some common mathematical fractals with structures comprising of <u>exact</u> copies of themselves at <u>all</u> magnifications. These objects possess **exact self-similarity** and are known as **regular fractals**. In chapter 1, a fractal object was loosely defined as one which appears self-similar at various scales of magnification and also as an object with its own **fractal dimension**, which is usually (but not always) a non-integer dimension greater than its topological dimension, D_T, and less than its Euclidean dimension, D_E. To date, there exists no watertight definition of a fractal object. Mandelbrot offered the following definition: 'A fractal is by definition a set for which the Hausdorff dimension strictly exceeds the topological dimension', which he later retracted and replaced with: 'A fractal is a shape made of parts similar to the whole in some way'. In this book, we will adopt, as a *test for a fractal object*, *the condition that its fractal dimension exceeds its topological dimension—whichever measure of fractal dimension is employed*. As we do this, bear in mind the ambiguous nature of the definition of a fractal.

2.2 The Cantor set

The **Cantor set** must certainly rank as one of the most frequently quoted fractal objects in the literature, alongside perhaps, the Koch curve and Mandelbrot set. It is arguably the simplest of fractals and a good place to begin our discussion on fractals and their geometric properties. The Cantor set consists of an infinite set of disappearing line segments in the unit interval. The best aid to the comprehension of the Cantor set fractal is an illustration of its method of construction. This is given in figure 2.1 for the simplest form of Cantor set, namely the triadic Cantor set. The set is generated by removing the middle third of the unit line segment (step $k = 1$ in the figure). From the two remaining line segments, each one third in length, the middle thirds are again removed (step $k = 2$ in the figure). The middle thirds of the remaining four line segments, each one-ninth in length, are then removed ($k = 3$) and so on to infinity. What is left is a collection of infinitely many disappearing line segments lying on the unit interval whose individual and combined lengths approach zero. This set of 'points' is known as a Cantor set, Cantor dust, or Cantor discontinuum.

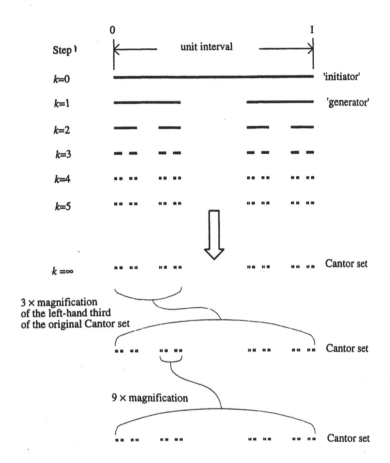

Figure 2.1. The construction of the triadic Cantor set.

In the construction of the Cantor set the initial unit line segment, $k = 0$, is known as the **initiator** of the set. The first step, $k = 1$, is known as the **generator** (or sometimes motif), as it is the repeated iteration of this step on subsequent line segments which leads to the generation of the set. Notice in the figure that the fifth iteration is indistinguishable from the Cantor set obtained at higher iterations. This problem occurs due to the limit of the finite detail our eyes (or the printer we use to plot the image) can resolve. Thus, to illustrate the set, it is sufficient to repeat the generation process only by the number of steps necessary to fool the eye, and not an infinite number of times. (This is true for all illustrations of fractal objects.) However, make no mistake, only after an infinite number of iterations do we obtain the Cantor set. For a finite number of iterations the object produced is merely a collection of line segments with finite measurable length. These objects formed *en route* to the fractal object are termed **prefractals**.

The Cantor set is a regular fractal object which exhibits exact self-similarity over all scales. This property is illustrated at the bottom of figure 2.1, where the left-hand third of the Cantor set is magnified three times. After magnification we see that the original

Cantor set is formed. Further zooming into one ninth of the newly formed set, we see that again the original set is formed. In fact copies of the Cantor set abound. Zooming into any apparent 'point' in the set produces the original set. It is easily seen that the Cantor set contains an infinite number of copies of itself, within itself, or to put it another way the Cantor set is made up of Cantor sets.

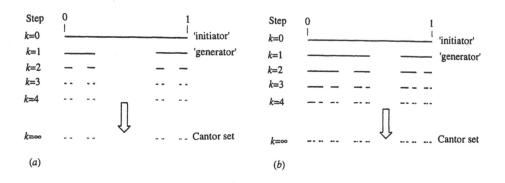

Figure 2.2. Two more examples of Cantor set construction. (*a*) Middle half removal. (*b*) Two-scale Cantor construction.

The triadic Cantor set described above is so called as it involves the removal of the middle third of the remaining line segments at each step in its construction. Any number of variants of the Cantor set may be formed by changing the form of the generator. Two such Cantor sets are shown in figure 2.2. The set on the left of the figure is formed by removing the middle half of each remaining line segment at each step, leaving the end quarters of the line. In the right-hand construction, the segment of each line removed at each stage leaves the first half of the original line and the last quarter. Again after an infinite number of steps a Cantor set is formed.

The Cantor set is simple in its construction, yet it is an object with infinitely rich structure. How do we make sense of the Cantor set? It does not fill up the unit interval continuously, as a line, i.e. one-dimensional object, nor is it a countable collection of zero-dimensional points. Rather, it fills up the unit interval in a special way and as a complete set has a dimension which is neither zero nor one, in fact it has a non-integer, fractal dimension somewhere in between zero and one. Non-integer, fractal dimensions are quite difficult to conceptualize initially and will be dealt with in the following sections.

2.3   Non-fractal dimensions: the Euclidean and topological dimensions

Generally, we can conceive of objects that are zero dimensional or 0D (points), 1D (lines), 2D (planes), and 3D (solids) see figure 2.3. We feel comfortable with zero, one, two and three dimensions. We form a 3D picture of our world by combining the 2D images from each of our eyes. Is it possible to comprehend higher-dimensional objects, i.e. 4D, 5D, 6D and so on? What about non-integer-dimensional objects such as 2.12D, 3.79D or 36.91232... D?

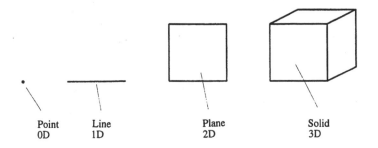

Point   Line   Plane   Solid
0D   1D   2D   3D

Figure 2.3. Common integer dimensions.

We will encounter many definitions of dimension as we proceed through this book. Before we deal with fractal dimensions, let us look at the two most common, and perhaps most comprehensible, definitions of dimension, the **Euclidean dimension**, D_E, and **topological dimension**, D_T. Both definitions lead to non-fractal, integer dimensions. The Euclidean dimension is simply the number of co-ordinates required to specify the object. The topological dimension is more involved. The branch of mathematics known as topology considers shape and form of objects from essentially a qualitative point of view. Topology deals with the ways in which objects may be distorted from one shape and formed into another without losing their essential features. Thus straight lines may be transformed into smooth curves or bent into 'crinkly' curves as shown in figure 2.4, where each of the constructions are topologically equivalent. Certain features are invariant under proper transformations (called homeomorphisms by topologists)— for instance, holes in objects remain holes regardless of the amount of stretching and twisting the object undergoes in its transformation from one shape to another. All of the two-holed surfaces in figure 2.5, although quite different in shape, are topologically equivalently as each one may be stretched and moulded into one of the others.

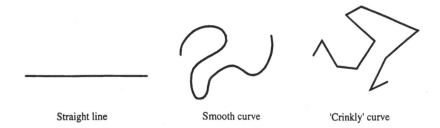

Straight line   Smooth curve   'Crinkly' curve

Figure 2.4. Topologically equivalent curves.

The topological dimension of an object does not change under the transformation of the object. The topological dimension derives from the ability to cover the object with discs of small radius. This is depicted in figure 2.6. The line segment may be covered using many discs intersecting many times with each other (figure 2.6(a)). However, it is possible to refine this covering using discs with only a single intersection between adjacent pairs of discs (figure 2.6(b)). Even when the line is contorted, one can find

Figure 2.5. Topologically equivalent forms—surfaces with two holes.

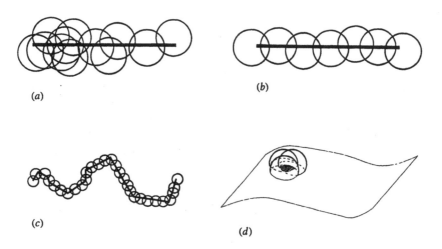

Figure 2.6. The covering of objects with discs and spheres to reveal the topological dimension. (*a*) Line segment covered by discs. (*b*) Line segment covered by discs only intersecting in pairs. (*c*) Crinkly line covered by discs only intersecting in pairs. (*d*) Surface covered by spheres (the intersection region is shaded.

discs sufficiently small to cover it with only intersections occurring between adjacent pairs of the covering discs, depicted in figure 2.6(*c*). The segment within each covering disc can itself be covered using smaller discs which require only to intersect in pairs. In a similar manner, a surface may be covered using spheres of small radius with a minimum number of intersections requiring intersecting triplets of spheres (figure 2.6(*d*)). The definition of the topological dimension stems from this observation. The covering of an object by elements (discs or spheres) of small radius requires intersections between a minimum of $D_T + 1$ groups of elements. Figure 2.7 shows a comprehensive set of common forms with their respective Euclidean and topological dimensions. Figure 2.8 contains the Cantor set. Its Euclidean dimension, D_E, is obviously equal to one, as we require one co-ordinate direction to specify all the points on the set. It can be seen from the figure that it is possible to find single non-intersecting discs of smaller and smaller radius to cover sub-elements of the set, thus the topological dimension, D_T, of the Cantor set is zero.

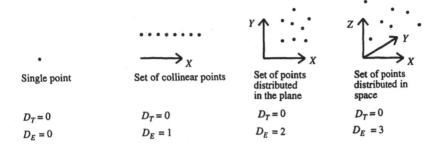

Single point	Set of collinear points	Set of points distributed in the plane	Set of points distributed in space
$D_T = 0$	$D_T = 0$	$D_T = 0$	$D_T = 0$
$D_E = 0$	$D_E = 1$	$D_E = 2$	$D_E = 3$

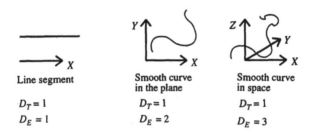

Line segment	Smooth curve in the plane	Smooth curve in space
$D_T = 1$	$D_T = 1$	$D_T = 1$
$D_E = 1$	$D_E = 2$	$D_E = 3$

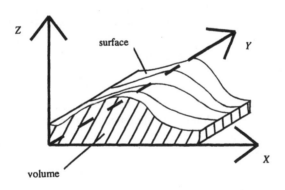

Section of solid object

$D_T(\text{surface}) = 2$ $D_T(\text{volume}) = 3$

$D_E(\text{surface}) = 3$ $D_E(\text{volume}) = 3$

Figure 2.7. A set of common forms with their respective Euclidean and topological dimensions.

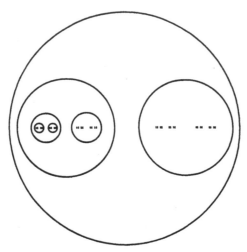

Figure 2.8. Covering the Cantor set with successively smaller, non-intersecting discs to reveal the topological dimension.

2.4 The similarity dimension

Their are many definitions of dimension which give a non-integer, or fractal, dimension. These dimensions are particularly useful in characterizing fractal objects.   In the remaining parts of this chapter we will concentrate on the **similarity dimension**, denoted D_S, to characterize the construction of regular fractal objects. As we proceed through subsequent chapters of the text further definitions of dimension will be introduced where appropriate.

The concept of dimension is closely associated with that of scaling. Consider the line, surface and solid depicted in figure 2.9, divided up respectively by self-similar sub-lengths, sub-areas and sub-volumes of side length ε. For simplicity in the following derivation assume that the length, L, area, A, and volume, V, are all equal to unity.

Consider first the line. If the line is divided into N smaller self-similar segments, each ε in length, then ε is in fact the scaling ratio, i.e. $\varepsilon/L = \varepsilon$, since $L = 1$. Thus

$$L = N\varepsilon = 1 \qquad (2.1a)$$

i.e. the unit line is composed of N self-similar parts scaled by $\varepsilon = 1/N$.

Now consider the unit area in figure 2.4. If we divide the area again into N segments each ε^2 in area, then

$$A = N\varepsilon^2 = 1 \qquad (2.1b)$$

i.e. the unit surface is composed of N self-similar parts scaled by $\varepsilon = 1/N^{1/2}$.

Applying similar logic, we obtain for a unit volume

$$V = N\varepsilon^3 = 1 \qquad (2.1c)$$

i.e. the unit solid is N self-similar parts scaled by $\varepsilon = 1/N^{1/3}$.

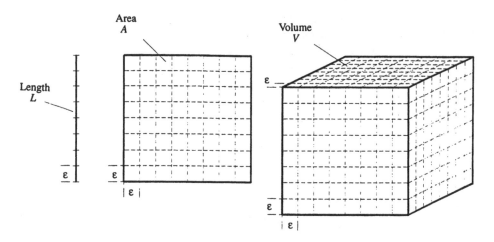

Figure 2.9. Scaling and dimension. Each object consists of N elements of side length ε, N is determined by the choice of ε. It should be noted that N for each object need not necessarily be the same, as is the case shown above.

Examining expressions (2.1a–c) we see that the exponent of ε in each case is a measure of the (similarity) dimension of the object, and we have in general

$$N\varepsilon^{D_S} = 1. \tag{2.2}$$

Using logarithms leads to the expression,

$$D_S = \frac{\log(N)}{\log(1/\varepsilon)}. \tag{2.3}$$

Note that here the subscript 'S' denotes the similarity dimension.

The above expression has been derived using familiar objects which have the same integer Euclidean, topological and similarity dimensions, i.e. a straight line, planar surface and solid object, where $D_E = D_S = D_T$. However, equation (2.3) may also be used to produce dimension estimates of fractal objects where D_S is non-integer. This can be seen by applying the above definition of the self-similar dimension to the triadic Cantor set constructed in section 2.2, (see figures 2.1 and 2.8). From figure 2.1 we saw that the left-hand third of the set contains an identical copy of the set. There are two such identical copies of the set contained within the set, thus $N = 2$ and $\varepsilon = \frac{1}{3}$. According to equation (2.3) the similarity dimension is then

$$D_S = \frac{\log(2)}{\log(1/(1/3))} = \frac{\log(2)}{\log(3)} = 0.6309\ldots. \tag{2.4a}$$

Thus, for the Cantor set, D_S is less than one and greater than zero: in fact it has a non-integer similarity dimension of 0.6309... due to the fractal structure of the object. We saw in the previous section that the Cantor set has Euclidean dimension of one and a topological dimension of zero, thus $D_E > D_S > D_T$. As the similarity dimension exceeds the topological dimension, according to our test for a fractal given in section 2.1, the set is a fractal with a fractal dimension defined by the similarity dimension of 0.6309.... As an aid to comprehension it may be useful to think of the Cantor set as neither a line nor a point, but rather something in between.

Instead of considering each sub-interval of the Cantor set scaled down by one-third we could have looked at each subinterval scaled down by one-ninth. As we saw from figure 2.1, there are four such segments, each an identical copy of the set. In this case $N = 4$ and $\varepsilon = \frac{1}{9}$ and again this leads to a similarity dimension of

$$D_S = \frac{\log(4)}{\log(1/(1/9))} = \frac{\log(4)}{\log(9)} = \frac{2\log(2)}{2\log(3)} = 0.6309\ldots. \qquad (2.4b)$$

Similarly there are eight smaller subintervals containing identical copies of the set each at a scale of $\frac{1}{27}$ of the original set, giving

$$D_S = \frac{\log(8)}{\log(1/(1/27))} = \frac{\log(8)}{\log(27)} = \frac{3\log(2)}{3\log(3)} = 0.6309\ldots \qquad (2.4c)$$

and so on.

By now a general scaling rule is apparent. The general expression for the similarity dimension of the Cantor set is

$$D_S = \frac{\log(2^C)}{\log(3^C)} = \frac{C\log(2)}{C\log(3)} = \frac{\log(2)}{\log(3)} = 0.6309\ldots \qquad (2.4d)$$

where the scaling constant, C, depends on the scale used to identify the self-similarity of the object. It can be seen from the above that the similarity dimension is independent of the scale used to investigate the object.

2.5   The Koch curve

The **Koch curve**, the method of construction of which is illustrated in figure 2.10, is another well documented fractal. As with the Cantor set, the Koch curve is simply constructed using an iterative procedure beginning with the initiator of the set as the unit line segment (step $k = 0$ in the figure). The unit line segment is divided into thirds, and the middle third removed. The middle third is then replaced with two equal segments, both one-third in length, which form an equilateral triangle (step $k = 1$): this step is the generator of the curve. At the next step ($k = 2$), the middle third is removed from each of the four segments and each is replaced with two equal segments as before. This process is repeated an infinite number of times to produce the Koch curve. Once again the self-similarity of the set is evident: each sub-segment is an exact replica of the original curve, as shown in figure 2.11.

A noticeable property of the Koch curve is that it is seemingly infinite in length. This may be seen from the construction process. At each step, k, in its generation, the length of the prefractal curve increases to $\frac{4}{3}L_{k-1}$, where L_{k-1} is the length of the curve in the proceeding step. As the number of generations increase the length of the curve diverges. It is therefore apparent that length is not a useful measure of the Koch curve, as

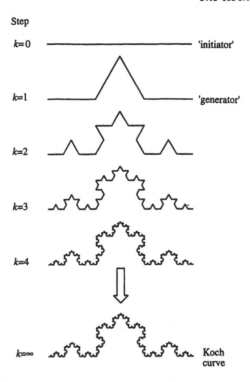

Figure 2.10. The construction of the Koch curve.

defined in the limit of an infinite number of iterations. In addition, it can be shown that the Koch curve is effectively constructed from corners, hence no unique tangent occurs anywhere upon it. The Koch curve is not a smooth curve and is nowhere-differentiable, as a unique tangent, or slope, cannot be found anywhere upon it.

The Koch curve is a fractal object possessing a fractal dimension. Each smaller segment of the Koch curve is an exact replica of the whole curve. As we can see from figure 2.11, at each scale there are four sub-segments making up the curve, each one a one third reduction of the original curve. Thus, $N = 4$, $\varepsilon = \frac{1}{3}$, and the similarity dimension based on expression (2.3) is

$$D_S = \frac{\log(N)}{\log(1/\varepsilon)} = \frac{\log(4)}{\log(3)} = 1.2618\ldots \tag{2.5}$$

that is, the Koch curve has a dimension greater than that of the unit line ($D_E = D_T = 1$) and less than that of the unit area ($D_E = D_T = 2$). The Euclidean dimension of the Koch curve, D_E, is two as we need two co-ordinate directions to specify all points on it. The topological dimension, D_T, of the Koch curve is unity, as we can cover it with successively smaller discs intersecting in pairs. The similarity dimension of the Koch curve lies between its Euclidean and topological dimension, i.e. $D_E > D_S > D_T$, which leads us to conclude that it is indeed a fractal object, with a fractal (similarity) dimension, D_S, of 1.2618.... .

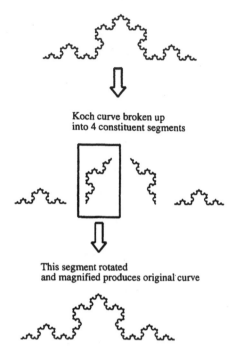

Koch curve broken up
into 4 constituent segments

This segment rotated
and magnified produces original curve

Figure 2.11. The self-similar structure of the Koch curve.

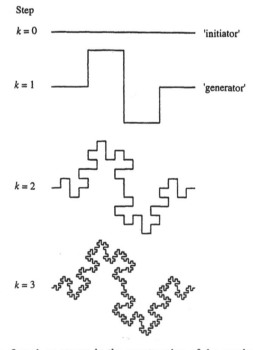

Step

$k = 0$ 'initiator'

$k = 1$ 'generator'

$k = 2$

$k = 3$

Figure 2.12. The first three stages in the construction of the quadratic Koch curve.

2.6   The quadratic Koch curve

The Koch curve shown in both figures 2.10 and 2.11 is more specifically known as the triadic Koch curve. As with the triadic Cantor set, the triadic Koch curve's name stems from the fact that the middle thirds of the line segments are modified at each step. By changing the form of the generator a variety of Koch curves may be produced. Figure 2.12 contains the first three stages in the construction of the quadratic Koch curve, also known as the Minkowski sausage. This curve is generated by repeatedly replacing each line segment, composed of four quarters, with the generator consisting of eight pieces, each one quarter long (see figure 2.12). As with the triadic Koch curve the Minkowski sausage is a fractal object. Each smaller segment of the curve is an exact replica of the whole curve. There are eight such segments making up the curve, each one a one-quarter reduction of the original curve. Thus, $N = 8$, $\varepsilon = \frac{1}{4}$, and the similarity dimension based on expression (2.3) is

$$D_S = \frac{\log(N)}{\log(1/\varepsilon)} = \frac{\log(8)}{\log(4)} = \frac{3\log(2)}{2\log(2)} = \frac{3}{2} = 1.5. \tag{2.6}$$

Figure 2.13 contains four more Koch curves produced using a variety of generators. The reader is invited to define his, or her, own generators, and use them to produce new fractal curves.

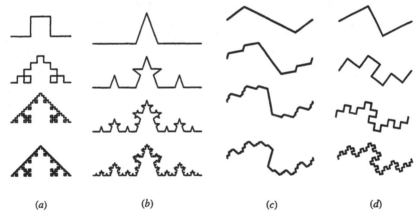

    (a)   (b)   (c)   (d)

Figure 2.13. Miscellaneous Koch curve constructions (all have unit line initiators—not shown).

2.7   The Koch island

The Koch island (or snowflake) is composed of three Koch curves rotated by suitable angles and fitted together: its construction is shown in figure 2.14. We already know that the length of the Koch curve is immeasurable, so the length of the coastline of the Koch island is seemingly infinite, but what about the area bounded by the perimeter of the island? It certainly looks finite. We can obtain a value for the bounded area by examining the construction process. Let us first assume for simplicity that the initiator is

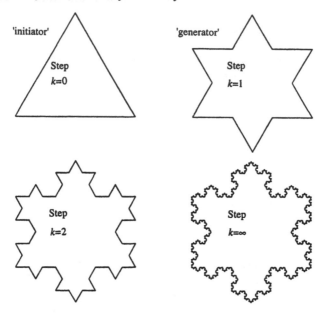

Figure 2.14. The Koch island and its construction.

composed of three unit lines. The area bounded by the perimeter is then half of the base multiplied by the height of the equilateral triangle, i.e. $\frac{1}{2} \times 1 \times \sqrt{3}/2$. At step $k = 1$ three smaller triangles are added, each with a base length equal to one third. At step $k = 2$ another twelve smaller triangles are added, each with base length equal to one ninth. At step $k = 3$ (not shown in the figure) forty-eight smaller triangles are added, each with base length of one twenty-seventh. The area then increases at each stage as follows:

$$\text{step } k = 0 \qquad \text{area } = \frac{1}{2} \times \frac{\sqrt{3}}{2} \times 1 = \frac{\sqrt{3}}{4} \tag{2.7a}$$

$$\text{step } k = 1 \qquad \text{area } = \frac{\sqrt{3}}{4} + 3\left(\frac{\sqrt{3}}{4 \times 3} \times \frac{1}{3}\right) \tag{2.7b}$$

$$\text{step } k = 2 \qquad \text{area } = \frac{\sqrt{3}}{4} + 3\left(\frac{\sqrt{3}}{4 \times 3} \times \frac{1}{3}\right) + 12\left(\frac{\sqrt{3}}{4 \times 9} \times \frac{1}{9}\right) \tag{2.7c}$$

$$\text{step } k = 3 \qquad \text{area } = \frac{\sqrt{3}}{4} + 3\left(\frac{\sqrt{3}}{4 \times 3} \times \frac{1}{3}\right)$$

$$+ 12\left(\frac{\sqrt{3}}{4 \times 9} \times \frac{1}{9}\right) + 48\left(\frac{\sqrt{3}}{4 \times 27} \times \frac{1}{27}\right) \tag{2.7d}$$

In general, for an arbitrary step k

$$\text{Area} = \frac{3 \times \sqrt{3}}{4 \times 4}\left(\frac{4}{3} + \frac{4^1}{9^1} + \frac{4^2}{9^2} + \frac{4^3}{9^3} + \ldots + \frac{4^k}{9^k}\right). \tag{2.7e}$$

We may then split up this expression to give

$$\text{area} = \frac{3 \times \sqrt{3}}{16} \left(\frac{1}{3}\right) + \frac{3 \times \sqrt{3}}{16} \left(1 + \frac{4^1}{9^1} + \frac{4^2}{9^2} + \frac{4^3}{9^3} + \cdots + \frac{4^k}{9^k}\right). \qquad (2.7f)$$

In the limit, as k tends to infinity, the geometric series in the brackets on the right-hand side of the above expression tends to $\frac{9}{5}$: this leaves us with an area of

$$\text{area} = \frac{3 \times \sqrt{3}}{16} \left(\frac{1}{3} + \frac{9}{5}\right) = \frac{3 \times \sqrt{3}}{16} \times \frac{32}{15} = \frac{2}{5}\sqrt{3}. \qquad (2.7g)$$

The Koch island therefore has a finite area of $\frac{2}{5}\sqrt{3}$, or about 0.693 units (of area). Thus, the Koch island has a regular area, in the sense that it is bounded and measurable, but an irregular, immeasurable perimeter. To generate the Koch island, we used three Koch curves with unit initiator. However, if the initiator were a in length, then the area would be simply $\frac{2}{5}\sqrt{3}a^2$. You can easily verify this for yourself. We will return briefly to the Koch island in our discussion of the fractal nature of natural coastlines in the next chapter.

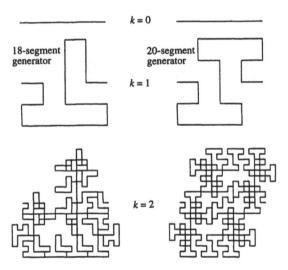

Figure 2.15. The construction process for curves with similarity dimension greater than two.

2.8 Curves in the plane with similarity dimension exceeding two

The similarity dimension can exceed two for curves in the plane. This may initially seem counter-intuitive, however, it may be easily demonstrated. Figure 2.15 contains two curves whose generators replace the original line segments with curves consisting of eighteen and twenty segments respectively, each of side length one quarter of the original. The similarity dimension of the curve resulting from the eighteen-segment generator is

$$D_S = \frac{\log(N)}{\log(1/\varepsilon)} = \frac{\log(18)}{\log(4)} = 2.0849\ldots. \qquad (2.8)$$

Similarly, $D_S = 2.1609\ldots$ for the twenty-segment curve. The similarity dimension exceeds two in these cases due to the overlapping parts of the fractal curve. Here, for both curves we have the slightly unusual condition that $D_S > D_E > D_T$, however, as the fractal dimension exceeds the topological dimension the object is still fractal by our definition. One way to avoid the fractal dimension exceeding D_E is to use alternative definitions of dimension which only count overlapping parts of the curve once. These will be explored in the next chapter.

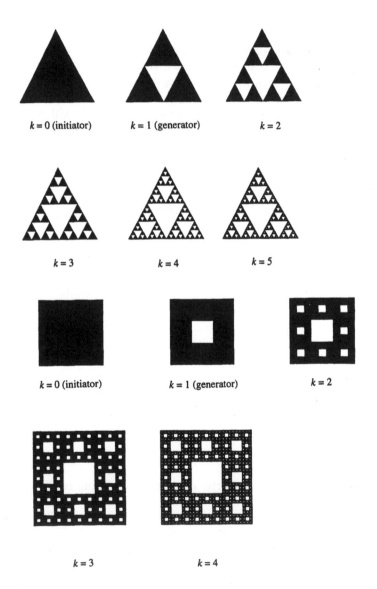

Figure 2.16. Construction of the Sierpinski gasket (top) and carpet (bottom).

2.9　The Sierpinski gasket and carpet

The construction of the **Sierpinski gasket** is illustrated in figure 2.16. The initiator in this case is a filled triangle in the plane. The middle triangular section is removed from the original triangle. Then the middle triangular sections are removed from the remaining triangular elements and so on. After infinite iterations the Sierpinski gasket is formed. Each prefractal stage in the construction is composed of three smaller copies of the preceding stage, each copy scaled by a factor of one half. The similarity dimension is given by

$$D_S = \frac{\log(N)}{\log(1/\varepsilon)} = \frac{\log(3)}{\log(2)} = 1.5849\ldots. \tag{2.9}$$

A sister curve to the Sierpinski gasket is the **Sierpinski carpet** also shown in figure 2.16. Its method of construction is similar to that of the gasket: this time the initiator is a square and the generator removes the middle square, side length one-third, of the original square. With both the Sierpinski gasket and carpet, the constructions lead to fractal curves whose area vanishes.

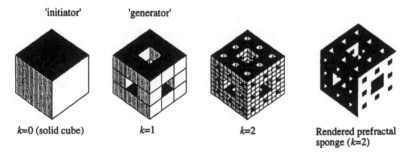

'initiator'　　'generator'

$k=0$ (solid cube)　　$k=1$　　$k=2$　　Rendered prefractal sponge ($k=2$)

Figure 2.17. Constructing the Menger sponge.

2.10　The Menger sponge

So far we have looked at constructions on the line (Cantor set) and in the plane (Koch curve and Sierpinski gasket and carpet). We end this chapter with an interesting object constructed in 3D space—the **Menger sponge**. Its construction is shown in figure 2.17 and, as can be seen, it is closely related to the Sierpinski carpet. The initiator in the construction is a cube. The first iteration towards the final fractal object, the generator, is formed by 'drilling through' the middle segment of each face. This leaves a prefractal composed of twenty smaller cubes each scaled down by one-third. These cubes are then drilled out leaving 400 cubes scaled down by one-ninth from the original cube (step $k = 2$ in the figure). Repeated iteration of this construction process leads to the Menger sponge. The similarity dimension of the Menger sponge is

$$D_S = \frac{\log(N)}{\log(1/\varepsilon)} = \frac{\log(20)}{\log(3)} = 2.7268\ldots \tag{2.10}$$

which is between its topological dimension of one (as it is a curve with zero volume, zero area and infinite length) and Euclidean dimension of three.

2.11 Chapter summary and further reading

2.11.1 Chapter keywords and key phrases

exact self-similarity	*regular fractals*	*fractal dimension*
Cantor set	*initiator*	*generator*
prefractals	*Euclidean dimension*	*topological dimension*
similarity dimension	*Koch curve*	*Sierpinski gasket/carpet*
Menger sponge		

2.11.2 Summary and further reading

In this chapter we have been introduced to regular fractal objects which have exact self-similarity at all scales, i.e. each small part of the object contains identical copies of the whole. To characterize these fractals requires that we re-evaluate our concepts of dimension. The Euclidean and topological definitions of dimension give only integer values. To obtain a fractal dimension for the exactly self-similar fractals we used, possibly the simplest definition of fractal dimension, the similarity dimension, D_S. In general, if the similarity dimension is greater than the topological dimension of the object then the object is a fractal, and, more often than not, the fractal dimension is a non-integer value. For more examples and information on exactly self-similar fractals the reader is referred for an elementary introduction to the book by Lauwerier (1991), and for an intermediate and comprehensive introduction to the book by Mandelbrot (1977), or, better, the extended version of this text by Mandelbrot (1982a). In-depth accounts of the Cantor set, Koch curve, Sierpinski gasket and Menger sponge, together with brief biographical details of their originators, are given by Peitgen *et al* (1992a). A method for the generation of the Sierpinski gasket using random numbers is given, amongst other useful information, by Peitgen *et al* (1991). Reiter (1994) presents some computer generated generalizations of the Sierpinski gaskets and carpets and the Menger sponge. The computer generation of the Koch curve is discussed by Hwang and Yang (1993). David (1995) presents two examples of 3D regular fractals based on Keplerian solids and Wicks (1991) presents an advanced mathematical account of fractals and hyperspaces.

Much of the interest in fractal geometry lies in its ability to describe many natural objects and processes, however, generally these are not exactly self-similar but rather statistically self-similar, whereby each small part of the fractal has the same statistical properties as the whole. We move on to these statistical, or random, fractals in the following chapter.

2.12 Revision questions and further tasks

Q2.1 List the keywords and key phrases of this chapter and briefly jot down your understanding of them.

Q2.2 Sketch out a line of unit length and plane of unit area. By selecting appropriate self-similar parts find their Euclidean, topological, and similarity dimensions, and show that they are not fractal objects.

Q2.3 (*a*) What are the Euclidean and topological dimensions of the Koch curve constructions shown in figure 2.13?

(*b*) Assuming that the line segments of the generator in figure 2.13(*a*) are all the same length, calculate the similarity dimension of the resulting fractal curve.

(*c*) The lengths of the line segments in the generator of figure 2.13(*d*) are each one half of the original unit line initiator. What is the similarity dimension of the resulting fractal curve?

(*d*) Try to produce some of your own Koch curves and find their similarity dimension.

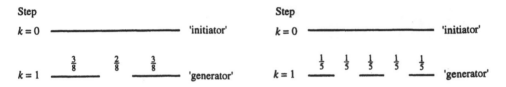

Figure Q2.4. The initiator and generator for two Cantor sets.

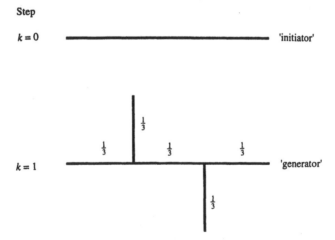

Figure Q2.5. The initiator and generator of a fractal curve.

Q2.4 The initiator and generator of two Cantor sets are given above in figure Q2.4. On graph paper generate the first four prefractals of the set. What is the similarity dimension of the resulting Cantor sets after infinite repetitions of the construction process?

Q2.5 The initiator and generator of a fractal curve are given in figure Q2.5. On graph paper generate the first few (as many as you can) prefractals of the set. What is the similarity dimension of the resulting fractal curve after infinite repetitions of the construction process?

Q2.6 (*a*) Show that the similarity dimension of the fractal curve on the right-hand side of figure 2.15 is greater than two.

(*b*) Explain why a curve generated in the plane can have a similarity dimension exceeding 2.

(*c*) Try to produce your own curves with $D_S > 2$.

Q2.7 What is the similarity dimension of the Sierpinski carpet shown in figure 2.16?

Figure Q2.8. The construction of a fractal dust.

Q2.8 The generation process of a fractal dust is given in figure Q2.8. At each stage the remaining squares are divided into sixteen smaller squares and twelve of them removed. What is the resulting self-similarity dimension of the dust? What is the topological and Euclidean dimension of the dust? Explain why it is a fractal object.

Q2.9 (*a*) Generate the $k = 3$ iteration of a quadratic Koch island. Use the unit square as the initiator and the construction given in figure 2.12 to generate the curve on each of the four initiators.

(*b*) What is the area of the quadratic Koch island?

(*c*) If the initiator were a square of side length a, what would the resulting area of the Koch island be?

Chapter 3

Random fractals

3.1 Introduction

In chapter 2 we investigated the properties of regular fractals—regular in the sense that they are composed of scaled down and rotated <u>identical</u> copies of themselves. The exact structure of regular fractals is repeated within each small fraction of the whole, i.e. they are exactly self-similar. There is, however, another group of fractals, known as **random fractals**, which contain a random or statistical element. These fractals are not exactly self-similar, but rather **statistically self-similar**. Each small part of a random fractal has the same statistical properties as the whole. Random fractals are particularly useful in describing the properties of many natural objects and processes.

3.2 Randomizing the Cantor set and Koch curve

A simple way to generate a fractal with an element of randomness is to add some probabilistic element to the construction process of a regular fractal, such as those investigated in chapter 2. A random version of the Cantor set may be produced in several ways, and figure 3.1 illustrates two methods for producing random Cantor sets. In the method depicted at the top of figure 3.1, the triadic Cantor set construction process is modified to allow the removal of any one third of each line segment at each prefractal stage. The third to be removed is selected at random. The generation of the set at the bottom of figure 3.1 again shows a random generation of a Cantor set, this time each remaining line segment is replaced with two smaller segments of random length. As we zoom into each of these Cantor sets we find that they have a statistical self-similarity as their construction involves the same random process at all scales.

Figure 3.2 illustrates the construction of a random version of the Koch curve. Its construction is very similar to that of the triadic Koch curve, investigated in chapter 2, in that the generation process involves the removal of the middle third of the remaining line segments at each prefractal stage, replacing the removed segment with two sides of an equilateral triangle. This time, however, the replacement elements are randomly placed either side of the removed segment. The resulting random fractal curve looks rather irregular compared to its cousin, the exactly self-similar triadic Koch curve of figure 2.10. As with its regular counterpart, any attempt to measure the length of the random

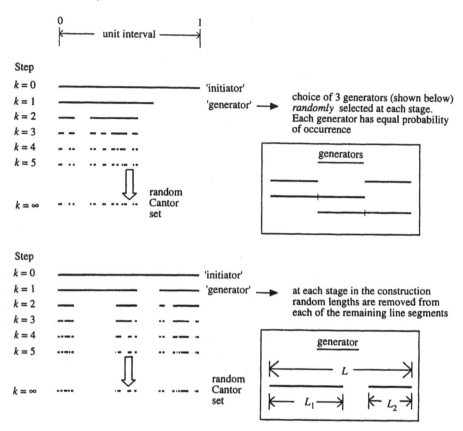

Figure 3.1. Two methods for randomizing the Cantor set.

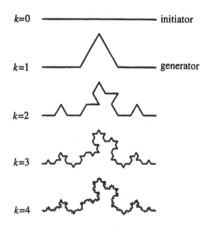

Figure 3.2. The first four steps in the construction of the randomized Koch curve.

Koch curve is futile as the measurement diverges with each iteration. However, the random Koch curve still retains a degree of regularity as, at each stage in the construction process, regular triangular features are placed randomly either side of the removed line segment.

3.3   Fractal boundaries

A fractal boundary is a non-crossing fractal curve which reveals more structure as one zooms in. The Koch island investigated in chapter 2 (figure 2.14) had a boundary, or coastline, consisting of three Koch curves, i.e. it had a regular fractal boundary. Real coastlines also tend to appear rather rugged at all levels of magnification: this is shown schematically in figure 3.3. However, unlike the coastline of the Koch island a real coastline is statistically self-similar in that randomly selected segments of the coastline possess the same statistical properties over all scales of magnification. Coastlines are, then, random fractal curves.

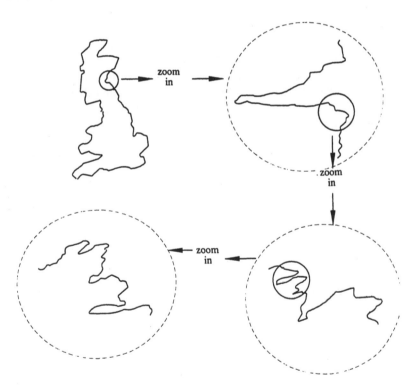

Figure 3.3. Zooming into a coastline to reveal more structure.

It is worthwhile, at this stage, to restate (see chapter 1) that even the best examples of natural fractals, including coastlines, possess self-similarity over a <u>finite</u> range of scales. This range is, however, sufficiently large to allow fractal geometry to play an important role in their characterization. We will come back to this point when discussing the typical regions of the Richardson plot later in the chapter.

The use of a dimension measurement is a good way to characterize and quantify the statistical self-similarity property of random fractal boundaries. We will not use the similarity dimension defined in chapter 2 as it relies on the identification of exactly self-similar parts at different scales of magnification to produce the dimension measurement. Random fractals do not possess exactly self-similar parts with which to indicate the scaling of the object, hence we need other methods to characterize their scaling properties. Two estimates of the fractal dimension of random coastline fractals which are commonly employed are the **box counting method** and the **structured walk technique**. These are examined in detail in the following sections.

3.4   The box counting dimension and the Hausdorff dimension

In chapter 2 we looked at the Euclidean and topological dimensions, both of which are integer dimensions. In addition, we used the similarity dimension to produce fractal dimensions for fractal objects. There are, however, many more definitions of dimension which produce fractal dimensions. One of the most important in classifying fractals is the **Hausdorff dimension**. In fact, Mandelbrot suggested that a fractal may be defined as an object which has a Hausdorff dimension which exceeds its topological dimension. A complete mathematical description of the Hausdorff dimension is outside the scope of this text. In addition, the Hausdorff dimension is not particularly useful to the engineer or scientist hoping to quantify a fractal object, the problem being that it is practically impossible to calculate it for real data. We therefore begin this section by concentrating on the closely related **box counting dimension** and its application to determining the fractal dimension of natural fractals before coming to a brief explanation of the Hausdorff dimension.

(i) *The box counting dimension*. To examine a suspected fractal object for its box counting dimension we cover the object with covering elements or 'boxes' of side length δ. The number of boxes, N, required to cover the object is related to δ through its box counting dimension, D_B. The method for determining D_B is illustrated in the simple example of figure 3.4, where a straight line (a one-dimensional object) of unit length is probed by cubes (3D objects) of side length δ. We require N cubes (volume δ^3) to cover the line. Similarly, if we had used squares of side length δ (area δ^2) or line segments (length δ^1), we would again have required N of them to cover the line. Equally, we could also have used 4D, 5D, or 6D elements to cover the line segment and still required just N of them. In fact, to cover the unit line segment, we may use any elements with dimension greater than or equal to the dimension of the line itself, namely one. To simplify matters, the line in figure 3.4 is specified as exactly one unit in length. The number of cubes, squares or line segments we require to cover this unit line is then $N\delta$ ($= 1$), hence $N = 1/\delta^1$. Notice that the exponent of δ remains equal to one regardless of the dimension of the probing elements, and is in fact the box counting dimension, D_B, of the object under investigation. Notice also that for the unit (straight) line $D_E = D_B = D_T$ ($= 1$), hence it is not a fractal by our definition in section 2.1, as the fractal dimension, here given by D_B, does not exceed the topological dimension, D_T.

To generalize the above and aid in the following discussion it is useful if, at this point, we rename all covering elements as hypercubes, as follows:

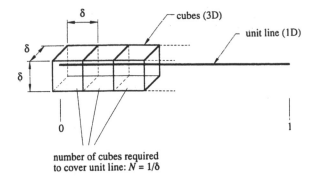

number of cubes required
to cover unit line: $N = 1/\delta$

Figure 3.4. A line (1D) probed using boxes (3D).

1D hypercube = 1D element, i.e. a line segment
2D hypercube = 2D element, i.e. a square
3D hypercube = 3D element, i.e. a box or cube
4D hypercube = 4D element
5D hypercube = 5D element
etc

and similarly rename all measurements as hypervolumes, V^*, as follows:

volume of 1D hypercube = 1D hypervolume (length)
volume of 2D hypercube = 2D hypervolume (area)
volume of 3D hypercube = 3D hypervolume (volume)
volume of 4D hypercube = 4D hypervolume
volume of 5D hypercube = 5D hypervolume
etc.

If we repeat the covering procedure outlined above for a plane unit area, it is easy to see that to cover such a unit area we would require $N = 1/\delta^2$ hypercubes of edge length δ and Euclidean dimension greater than or equal to two. Similarly, with a 3D solid object we would require $N = 1/\delta^3$ hypercubes of edge length δ with Euclidean dimension greater than or equal to three to cover it. Again notice that in each case the exponent of δ is a measure of the dimension of the object. In general, we require $N = 1/\delta^{D_B}$ boxes to cover an object where the exponent D_B is the box counting dimension of the object. We arrive at the following general formulation of D_B for objects of unit hypervolume:

$$D_B = \frac{\log(N)}{\log(1/\delta)} \tag{3.1}$$

obtained by covering the object with N hypercubes of side length δ. Note that the above expression is of rather limited use. It assumes the object is of unit hypervolume and in general will produce erroneous results for large δ. More general and practically useful expressions are given below in equations (3.4a, b). Note also the marked resemblance of equation (3.1) to the definition of the similarity dimension D_S given in equation (2.3).

However, do not confuse the two: the calculation of D_S requires that exactly self-similar parts of the fractal are identified, whereas D_B requires the object to be covered with self-similar boxes. Hence, D_B allows us greater flexibility in the type of fractal object that may be investigated.

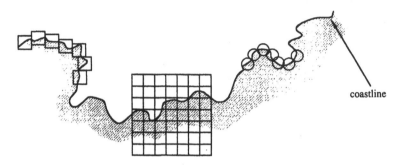

Figure 3.5. Determining the fractal dimension of a coastline using the box counting method.

The general expression for the dimension of an object with a hypervolume (i.e. length, area, volume or fractal hypervolume) not equal to unity, but rather given by V^*, is

$$D_B = \frac{\log (N) - \log(V^*)}{\log(1/\delta)} \tag{3.2}$$

where N is the number of hypercubes of side length δ required to cover the object, i.e. $N = V^*/\delta^{D_B}$. Rearranging equation (3.2) gives

$$\log (N) = D_B \log (1/\delta) + \log(V^*) \tag{3.3}$$

which is in the form of the equation of a straight line where the gradient of the line, D_B, is the box counting dimension of the object. This form is suitable for determining the box counting dimension of a wide variety of fractal objects by plotting $\log(N)$ against $\log(1/\delta)$ for probing elements of various side lengths, δ. Figure 3.5 illustrates three popular methods of covering a coastline curve using boxes and circles to obtain a box counting dimension estimate. One may place boxes against each other to obtain the minimum number required to cover the curve. Alternatively, one may use a regular grid of boxes and count the number of boxes, N, which contain a part of the curve for each box side length δ. Circles of diameter δ may also be used as probing elements to cover the curve, placing them so that they produce the minimum covering of the curve. In this case, δ corresponds to the diameter of the covering circles. Whichever method is used, we obtain the box counting dimension from the limiting gradient (as δ tends to zero) of a plot of $\log(N)$ against $\log(1/\delta)$, i.e. the derivative

$$D_B = \lim_{\delta \to 0} \frac{d (\log(N))}{d((\log(1/\delta))}. \tag{3.4a}$$

This is shown schematically in figure 3.6. In practice, the box counting dimension may be estimated by selecting two sets of $[\log(1/\delta) \log(N)]$ co-ordinates at small values of δ

(i.e. large values of $\log(1/\delta)$). An estimate of D_B is then given by

$$D_B = \frac{\log(N_2) - \log(N_1)}{\log(1/\delta_2) - \log(1/\delta_1)} \qquad (3.4b)$$

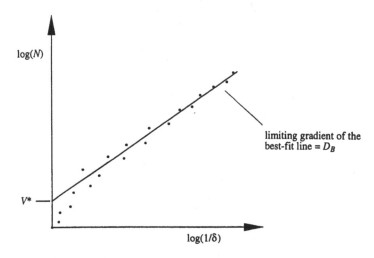

Figure 3.6. Estimating the box counting dimension of experimental data.

Alternatively, a more refined estimate may be obtained by drawing a best fit line through the points at small values of δ and calculating the slope of this line (see figure 3.6). For this case the N and δ values of equation (3.4b) are taken from two points on the best-fit line. This is particularly advisable where the data fluctuate at the limits of resolution.

The box counting dimension is widely used in practice for estimating the dimension of a variety of fractal objects. The technique is not confined to estimating the dimensions of objects in the plane, such as the coastline curve. It may be extended to probe fractal objects of high fractal dimension in multi-dimensional spaces, using multi-dimensional covering hypercubes. Its popularity stems from the relative ease by which it may be incorporated into computer algorithms for numerical investigations of fractal data. The grid method (central method depicted in figure 3.5) lends itself particularly to encoding within a computer program. By covering the data with grids of different box side lengths, δ, and counting the number of boxes, N, that contain the data, the box counting dimension is easily computed using equation (3.4b). We shall return to the box counting dimension in chapter 7 where we shall investigate its usefulness in characterizing fractal properties of structures associated with chaotic motion in multi-dimensional phase spaces.

(ii) *Hausdorff dimension.* There is a marked similarity between the box counting dimension and the Hausdorff dimension. Both use elements to cover the object under inspection. The difference between the two is really this. When using the box counting dimension we are asking, 'How many boxes or hypercubes do we need to <u>cover</u> the object?': in this case we only need use probing elements, or hypercubes, which have an integer dimension equal to or exceeding that of the object. In contrast, when using the Hausdorff dimension we are asking instead, 'What is the 'size' of the object?', that is we are trying to <u>measure</u> it. To measure its size or hypervolume we need to use the appropriate dimension of covering hypercubes, this appropriate dimension being the Hausdorff dimension D_H. In the rest of this section a brief overview of the Hausdorff dimension is given for completeness of the text.

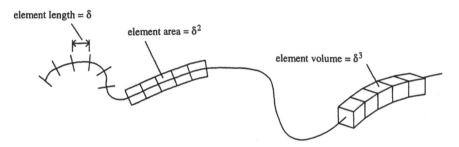

Figure 3.7. Measuring a smooth curve.

Let us first consider regular, non-fractal, or Euclidean objects. If we want to measure the size of an object we must use a measurement appropriate to its dimension, i.e. length is the natural measure of a line, area is the natural measure of a surface, and volume is the natural measure of a solid. Consider first the measurement of the smooth curve in figure 3.7. To measure the curve we can cover it with measuring elements such as lines, squares or cubes of linear size, δ, as shown in the figure. The length, area and volume of these measuring elements is given by δ^1, δ^2, δ^3 respectively, or more generally given by δ^{D_E}, where D_E is the integer Euclidean dimension of the measuring elements.

As with the box counting dimension we require N elements of side length δ to cover the curve. The measured length of the line, as measured by the covering elements, is given by

$$L_m = N\delta^1. \tag{3.5a}$$

As δ goes to zero in the limit, the measured length, L_m, tends to the true length of the curve L, i.e.

$$L_m \xrightarrow[\delta\to0]{} L. \tag{3.5b}$$

Now consider the covering of the curve using the square elements depicted in figure 3.7, each δ^2 in area. The measured area, A_m, associated with the line is then

$$A_m = N\delta^2; \tag{3.6a}$$

however, as δ tends to zero so does the measured area associated with the curve, i.e.

$$A_m \xrightarrow[\delta\to0]{} 0. \tag{3.6b}$$

This makes sense, as we expect a curve to have zero area. Similarly the measured volume, V_m, associated with the curve tends to zero as δ tends to zero, i.e.

$$V_m = N\delta^3 \tag{3.7a}$$

$$N\delta^3 \xrightarrow[\delta\to0]{} 0. \tag{3.7b}$$

Extending the above to higher-dimensional measuring elements, we see that only the length measure of a smooth curve gives a finite, non-zero answer equal to the length of the curve, L.

If we require N hypercubes to cover an object then the measured hypervolume V_m^* is given by the number of hypercubes multiplied by the volume of each hypercube, i.e.

$$V_m^* = N\delta^{D_E}. \tag{3.8a}$$

In the limit, as δ tends to zero, the measured hypervolume tends asymptotically to the actual hypervolume of the object, and is independent of δ, i.e.

$$V_m^* \xrightarrow[\delta \to 0]{} V^*. \tag{3.8b}$$

However, as we saw from the measurement of the smooth curve above, the measurement

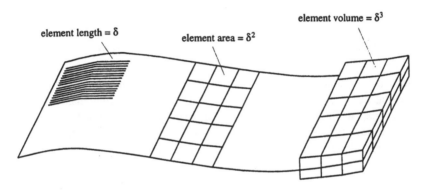

element length = δ element area = δ^2 element volume = δ^3

Figure 3.8. Measuring a smooth surface.

obtained is only sensible as long as the dimension of the measuring elements and the object are the same. Alternatively, if we attempt to measure an object with an inappropriate measuring element then we find that the measured hypervolume either tends to infinity, for measuring elements (hypercubes) of dimension less than that of the object, or zero, if we use measuring elements of dimension greater than the object. As another example, consider trying to find the length, area or volume of a surface using line segments, squares or cubes, as shown in figure 3.8. We cannot cover a surface with a finite number of lines, thus the measured hypervolume, i.e. the measured 'length', of the surface diverges as the length of the measuring line segments, δ, tends to zero, i.e.

$$V_m^* = N\delta \tag{3.9a}$$

$$N\delta \xrightarrow[\delta \to 0]{} \infty. \tag{3.9b}$$

Here V_m^* is the one-dimensional hypervolume, i.e. length. Similarly if we attempt to cover a surface with cubes of volume δ^3 we find that the measured volume tends to zero as δ tends to zero, i.e.

$$V_m^* = N\delta^3 \tag{3.10a}$$

$$N\delta^3 \xrightarrow[\delta \to 0]{} 0 \tag{3.10b}$$

and in fact for all measuring hypercubes of dimension greater than three, the hypervolume tends to zero as δ tends to zero. Only when we measure the surface using 2D hypercubes, i.e. squares, do we find that the measured hypervolume tends to a finite value in the limit, i.e.

$$V_m^* = N\delta^2 \tag{3.11a}$$

$$N\delta^2 \xrightarrow[\delta \to 0]{} V^* \tag{3.11b}$$

where V^* is in fact the area of the surface, A.

We see from the above that the measured hypervolume depends critically on the dimension of the probing hypercubes used. Generally the measured hypervolume is either zero or infinity, the change from zero to infinity occurring when the appropriate dimension of the measuring hypercube is used. The Hausdorff dimension is based upon the above observation: in its definition we are allowed to consider hypercubes, or test functions, with hypervolumes δ^D where the exponent D is non-integer. The Hausdorff dimension D_H of the object is defined as the critical dimension, D, for which the measured hypervolume changes from zero to infinity.

The Hausdorff and box counting dimensions enable non-integer dimensions to be found for fractal curves. Take for example the Koch curve which has a topological dimension of one, a Euclidean dimension of two, and a similarity dimension of 1.2618.... We saw before that length is not a useful measure for the Koch curve since the measured length of the prefractal curve diverges as one iterates the generation process. If we tried to measure it with line elements one would find that its measured length would tend to infinity as we used smaller and smaller line segments. Its Hausdorff dimension is therefore greater than one. If instead we probed the Koch curve with square areas of side length δ we would find that its measured area tends to zero as δ tends to zero. Thus, the Hausdorff dimension is less than two. In fact, the Koch curve has a Hausdorff and box counting dimension equal to its similarity dimension of 1.2618....

In practice, it is not possible to probe objects with non-integer hypercubes, hence the Hausdorff dimension estimate is not useful for determining the fractal dimension of real objects. The box counting dimension is closely related to the Hausdorff dimension and in most cases both produce the same fractal dimension estimate. In addition, both the Hausdorff and box counting dimension will often produce the same dimension estimate as the similarity dimension. However, problems occur with self-crossing curves as the similarity dimension takes account of the multiple layering of the self-crossing curve (as we saw in chapter 2) whereas the Hausdorff and box counting dimensions do not.

3.5   The structured walk technique and the divider dimension

A commonly used method for determining a fractal dimension estimate of fractal curves in the plane is the structured walk technique, illustrated in figure 3.9. The technique is much faster to perform by hand than the box counting dimension and requires the use of a compass or a set of dividers. (A ruler may also be used if neither of the first two pieces of equipment is available, but this does result in a more laborious task.) The method is outlined as follows.

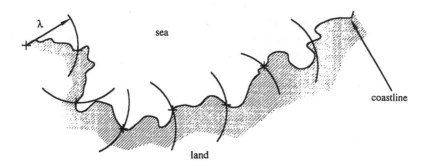

Figure 3.9. Determining the fractal dimension of a coastline using the structured walk technique.

(i)   Set the compass/dividers at a step length λ.
(ii) Take the initial point at the beginning of the curve (or select a suitable starting position if it is a closed curve).
(iii) Draw an arc, centred at the initial point, which crosses the curve.
(iv) The point where the arc <u>first</u> crosses the curve becomes the centre of the next arc.
(v)   Draw the next arc centred at the crossing point of step (iv).

Repeat steps (iv) and (v) until the end of the curve is reached.

(vi) Plot $\log(L)$ versus $\log(\lambda)$, where L is the length of the coastline measured using λ as a step length, i.e. $L = N\lambda$, where N is the number of steps taken to 'walk' along the curve.

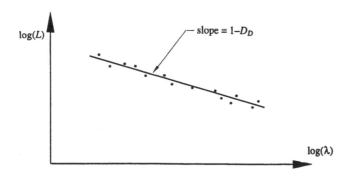

Figure 3.10. Richardson plot obtained from the structured walk technique.

(vii) Repeat steps (i) to (vi) for many step lengths, each time plotting $\log(L)$ versus $\log(\delta)$. The resulting plot is known as a **Richardson plot** (figure 3.10).
(viii) The slope, S, of the resulting curve is related to the **divider dimension**, D_D, by the relationship

$$S = 1 - D_D \qquad (3.12)$$

Hence, the dimension of the curve may be found by measuring S from the best fit line of the plotted points of step (vii). The slope of the Richardson plot is negative, i.e. the best fit line falls from left to right, thus $D_D > 1$. Note that when drawing successive arcs (steps (iv) and (v)) one may obtain slightly different dimension estimates depending upon the direction of approach of the arc. One may repeatedly swing clockwise into the coastline from the 'sea' (the inswing method), anti-clockwise out of the coastline from the 'land' (the outswing method), or alternate between the two (the alternate method). It is good practice to try all three methods and, in addition, to use various starting locations on the curve.

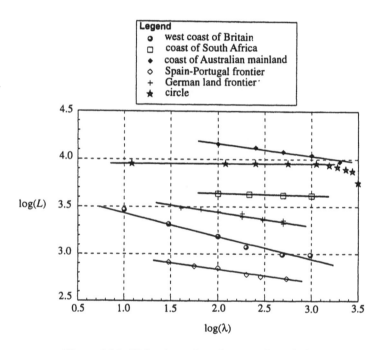

Figure 3.11. Richardson plot of country boundaries.

Figure 3.11 contains a Richardson plot of original data by L F Richardson who noted that reported lengths of the border between two countries were often claimed to be different by the two countries involved. For example, he noted that the Spanish–Portuguese border was stated as being 987 km and 1214 km by Spain and Portugal respectively, and similarly the Dutch–Belgian border was stated as being 380 km and 449 km respectively by the two countries. After investigation he reasoned that the differences could be attributed to the length of the measuring stick used in the calculation of the boundary length. The smaller the measuring stick length, λ, the longer the measured length L was found to be. On the Richardson plot of figure 3.11 the data points for a circle are also plotted. Notice that the circle boundary slope tends to zero for small values of λ, as the divider dimension, D_D, tends to the topological dimension, D_T ($= 1$). This implies that the circle boundary is not a fractal, and more, in that it is a smooth curve with measurable length.

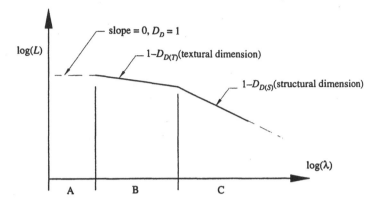

Figure 3.12. Main regions of the Richardson plot.

Quite often there is more than one distinct slope to be found on a Richardson plot (typically three). Figure 3.12 shows the typical slopes that may be found on such a plot. From the figure, three distinct slopes may be seen: these correspond to three distinct regions of the object under investigation. These regions are summarized as follows.

(i) *Region A*: λ is very small. A natural fractal may not be self-similar below this scale or alternatively the resolution of the fractal may not be sufficient to allow investigation below these scales. Thus, the curve appears smooth at these levels of magnification and the dimension tends to unity.

(ii) *Region B*: Small λ. In this region we are measuring the fine scale structure, or texture, of the curve. This gives the **textural dimension**—'$D_{D(T)}$'.

(iii) *Region C*: Larger λ. Here we are now measuring the larger scale structure of the curve. This gives the **structural dimension**—'$D_{D(S)}$'.

Regions B and C correspond to two different fractals, intertwined with each other to form the curve. Objects with two or more fractal dimensions are known as **multifractals**.

One area where the divider dimension has been applied with particular success is as a characterization tool in the classification of the fractal boundaries of fine particles such as soots, powders and dusts. (See the notes at the end of this chapter.) Figure 3.13(*a*) contains a simulated soot particle made up of circles connected tangentially. The outer boundary of the particle reproduces the general features of profiles typically found in agglomerations of soot particles from exhaust emissions. Figure 3.13(*b*) contains the Richardson plot of the particle boundary in figure 3.13(*a*). Notice the two distinct slopes associated with the structure and texture of the particle.

The divider dimension is therefore an extremely useful tool. However, its main shortcoming is that it is limited to the investigation of curves in the plane. Should we wish to measure fractal objects other than curves, say for example the surface of a cloud or fractal landscape, the box counting dimension should be used; this, due to its versatility, may be used to probe all manner of fractal objects occurring in multi-dimensional spaces.

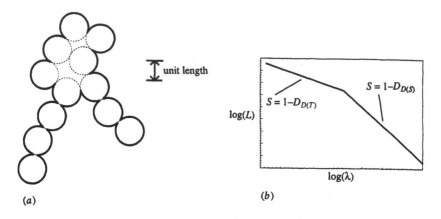

Figure 3.13. Richardson plot of a synthetic particle boundary. (*a*) Synthetic particle. (*b*) Richardson plot.

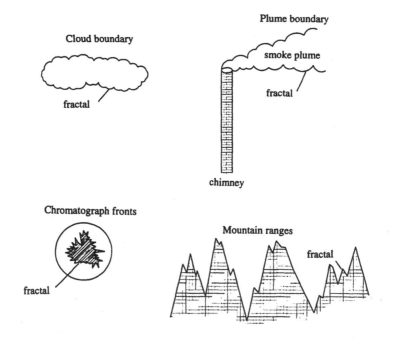

Figure 3.14. A selection of natural fractal objects.

Both the divider and box counting dimensions have been used to measure the fractal dimension of many natural fractals, including of course real coastlines and fine-particle boundaries, other examples include (see figure 3.14) cloud boundaries, smoke plume boundaries, chromatograph diffusion fronts, landscape profiles, and so on. Both dimension estimates are also useful in the estimation of the fractal dimension of crossing curves, such as fBm, which we shall encounter in chapter 4.

We leave this section by looking at the relationship between the box counting dimension and divider dimension on a fractal curve. First, we consider the box counting dimension. Rearranging equation (3.2) for non-unit hypervolumes we obtain

$$N(\delta)\delta^D = V^* \tag{3.13a}$$

where D is the box counting dimension. As an aid to clarity in the following discussion, we omit the subscript B of the box counting dimension and include δ in parenthesis to denote that N is a function of the box size δ. Hence, the number of boxes counted scales with δ as follows:

$$N(\delta) \propto \frac{1}{\delta^D} \tag{3.13b}$$

Considering now the structured walk technique, the measured length of the coastline, L, is a function of the divider length, λ, used, i.e.

$$N(\lambda)\lambda = L(\lambda). \tag{3.14}$$

Again, parentheses are used to denote that here the number of steps N required to walk along the curve, and hence the measured length L, is a function of step length λ. The linear scales in both techniques, δ and λ, are proportional to each other, i.e.

$$\lambda \propto \delta \tag{3.15a}$$

hence

$$N(\lambda) \propto N(\delta) \tag{3.15b}$$

Combining the above we obtain

$$\frac{1}{\lambda^D}\lambda \propto L(\lambda) \tag{3.16a}$$

or, more simply,

$$\lambda^{1-D} \propto L(\lambda). \tag{3.16b}$$

Hence, a log–log plot of λ against L results in a line with slope S equal to $1 - D$. Here we can see that the box counting dimension is in fact equivalent to the divider dimension for coastline curves. It is important to note, however, that this relationship between D_D and D_B does not hold for a special class of fractals with anisotropic scaling. We will come across these fractals, known as self-affine fractals, in the following chapter.

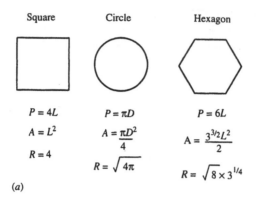

(a)

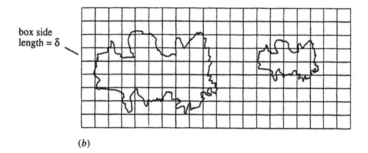

(b)

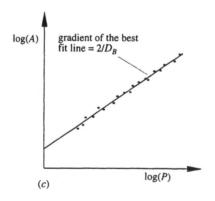

(c)

Figure 3.15. The perimeter–area relationship for similarly shaped bounded fractal islands. (a) Perimeter–area ratios, R, for common Euclidean shapes: L is the side length, D is the diameter (R is constant regardless of size of shape). (b) Similar islands with fractal boundaries (note that δ must be small enough to accurately measure the area of the smallest island). (c) A logarithmic plot of A against P revealing fractal dimension and self-similarity.

3.6 The perimeter–area relationship

We leave this chapter by briefly looking at an important relationship between an enclosed area, A, and its boundary perimeter, P, known as the **perimeter–area relationship**. For regular Euclidean shapes, i.e. circles, squares, triangles, hexagons, and so on, the ratio of the perimeter to the square root of enclosed area, R, is a constant regardless of the size of the shape, i.e.

$$R = \frac{P}{\sqrt{A}}. \tag{3.17a}$$

for a square, circle and hexagon, R is 4, $\sqrt{4\pi}$ and $\sqrt{8} \times 3^{1/4}$ respectively (see figure 3.15(a)). We can generalize this rule for areas bounded by fractal curves, where the length of the perimeter diverges as we use smaller and smaller measuring sticks to measure it. To do this we modify equation (3.17a) to

$$R = \frac{P^{1/D_B}}{\sqrt{A}} \tag{3.17b}$$

where P and A are now the measured perimeter and area of the boundary, using a length scale δ.

Equation (3.17b) is useful if, for example, we have a set of fractal shapes and we want to know whether they are statistically similar. Figure 3.15(b) shows a sketch of two random fractals islands at different scales. If these random fractals are the same shape then R given by equation (3.17b) should hold. In practice, we can investigate a whole series of fractal island shapes by covering them with a grid of box size δ. Then $P = N_p \delta$ and $A = N_a \delta$ where N_p and N_a are the numbers of boxes required to cover the perimeter and bounded area respectively. As long as we choose δ small enough to accurately measure the area of the smallest fractal island then a logarithmic plot of A against P for each island should produce a slope equal to $2/D_B$. This is illustrated in figure 3.15(c). In this way we can not only find the fractal dimension of the boundaries of a group of fractal islands, but also check that they are of similar shape. This technique has been used successfully to classify many sets of fractal shapes including rain clouds, fracture surfaces, contours and lake perimeters.

3.7 Chapter summary and further reading

3.7.1 Chapter keywords and key phrases

random fractals	*statistical self-similarity*	*Hausdorff dimension*
box counting method	*box counting dimension*	*structured walk technique*
Richardson plot	*divider dimension*	*structural dimension*
textural dimension	*multifractals*	*perimeter–area relationship*

3.7.2 General

In this chapter we investigated fractal objects which are not exactly self-similar in structure (as were those encountered in chapter 2), but rather statistically self-similar, with each small part of the fractal replicating the statistical properties of the whole. To determine the fractal dimension of these random fractals we may use the box counting dimension or the divider dimension (curves only). The similarity dimension used in chapter 2 cannot be used for random fractals as it relies on exact self-similarity. As with chapter 2, the reader is referred in the first instance to the excellent book, *The Fractal Geometry of Nature*, by Benoit Mandelbrot (1982a), for more information on the fractal geometry of random fractals. In addition, the text contains a wide ranging historical review. The book by Briggs (1992) contains many beautiful photographs of natural fractal phenomena. Another book by Hirst and Mandelbrot (1994) contains a series of striking black and white photographs used to illustrate the fractal geometry of landscapes. Mainieri (1993) gives criteria for the equality of the Hausdorff and box counting dimensions. More advanced mathematical accounts of fractal geometry are to be found in the texts by Falconer (1985, 1990), Edgar (1990), Wicks (1991), Dobrushin and Kosuka (1993), Massopust (1994), and Mattila (1995). Fractal curves are comprehensively dealt with by Tricot (1995).

The motivation behind much of the interest in fractal geometry lies in its ability to characterize natural phenomena (Kadanoff, 1986). It is now realized that there are a large number of objects and processes found in nature which may be described in terms of their fractal properties. In general, these phenomena exhibit statistical self-similarity over a large but finite range of scales. In the remainder of this section, many examples are given of naturally occurring objects and processes which have been described in terms of their fractal properties, together with references to allow the reader to delve more deeply into his or her own particular field of interest. In most cases, only essential references are given for the reader to use as a starting point for a more specific search. Note that there is some repetition of topics as many texts contain a wide range of fractal examples.

3.7.3 Miscellaneous subject areas

Harrison (1995) provides an elementary introduction to **fractals in chemistry**. A general background on **fractal curves** and **fractal geometry,** as well as applications in chemistry, is given by Fan *et al* (1991): this monograph contains three detailed examples of the use of fractal concepts in real **chemical studies**. A comprehensive account of the uses of fractal geometry in heterogeneous chemistry is given in a collection of papers edited by Avnir (1989): the text illustrates many uses of fractal geometry to describe chemical substances and processes including the following: **polymers, aggregations, interfaces, electrodes, molecular diffusion, molecule surface interactions, reaction kinetics, adsorption, flow in porous media, chromatography, geochemistry** and **analysis of proteins.** Three papers by Coppens and Froment (1994, 1995a, b) detail diffusion and reaction in a **fractal catalyst.** An excellent collection of papers concerning **fractals in physics** is edited by Aharony and Feder (1990), and many of the papers are referenced individually within this text. Pietronero and Tossatti (1986) have also edited a collection of papers concerning the use of **fractal geometry in physics**. This text covers a wide variety of topics including **viscous fingering, cracked metals, diffusion fronts** and the **fractal structure of clouds**. A comprehensive treatise on **fractal geometry and surface growth** is given by Barabási and Stanley (1995) (see DLA fractals in chapter 4). Fractals in **biology and medicine** are treated in a series of articles edited by Nonnenmacher *et al* (1994). The role of fractal geometry in **physiology** is outlined by West (1990) in the first half of his book (see also West and Goldberger 1987). The potential applications of fractals in **electrical engineering** are given by Jacquin (1994), and in a special section concerning the topic in the proceedings of the IEEE (1993). This latter publication contains the following topics: **wavelets and fractals, a review of fractal image coding, fractals in circuit output, computer interconnection, electrical processes in fractal geometry, radar imagery** and **the ultrasonic characterization of fractal media**.

Kaye (1994) uses fractal geometry to describe many of the properties of fractals in engineering, including: **fine particle boundaries, filter geometry, fractures, powder mixing, concrete, smoke plume boundaries** and **percolation**. This extensive text contains many real and synthetic particles (and other forms) together with their Richardson plots. (The **original Richardson plot** is to be found in the article by Richardson (1961).) In addition, the text includes an interesting chapter containing thoughts on the use of fractal geometry in the natural sciences. A review of the measurement of boundary fractal dimensions of highly irregular objects is given by Allen *et al* (1995). The reader interested in the use of fractal geometry in geology and geophysics is referred to the book by Turcotte (1992), who covers many diverse areas including **geological fragmentation**, the **fractal statistics of earthquakes**, the **fractal grading of ore deposits** and the **fractal structure of topographic images**. The structure of **soil fabric** has been described in terms of fractals by Bartoli *et al* (1991) and Young and Crawford (1991), and modelled using fractals by Moore and Krepfl (1991). Architects should see the book by Bovill (1996) for an introduction to fractal geometry in **architecture and design**.

Two excellent texts edited by Bunde and Havlin give many examples of the use of fractal geometry in science. The texts include coverage of the following topics: **cracks and fractures, dielectric breakdown, viscous fingering, smoke particle aggregates, chromatograph fronts, diffusion fronts** (Bunde and Havlin 1991); **neurons, lung (respiratory tree), polymer structure, DNA, chemical reactions, flow through porous media** (Bunde and Havlin 1994). The text edited by Cherbit (1991) treats the following topics: **galactic clusters, manganese oxide, electrochemistry, renal filtration, culture growth, lung structure, diffusion fronts** and **fractal dimension relative to observer**. A brief discussion of the fractal properties of **lightning, fluid turbulence** and **crystal growth** is given in the book by Schroeder (1991). Another text, edited by Fleischmann *et al* (1989), deals with fractals in the natural sciences and contains papers on, amongst other things, **fractals and phase transitions, structure of fractal solids, fractal aggregates, electrodeposition, flow through porous media** and **fractal adsorption**.

Other areas which have been investigated for their fractal properties are (in no particular order) the following: **the trajectories of drifters on the ocean surface**, (Osborne *et al* 1989) (see also chapter 4); **stock market indicators** (Huang and Yang 1995); **moon crater distribution** (Peitgen and Saupe 1988); the **structure of dispersing plumes** (Sykes and Gabruk 1994); **hydraulic roughness variations** (Vieux and Farajalla

1994); **fluid turbulence** (Sreenivasan 1991); **vegetative ecosystems** (Hastings and Sugihara 1993); **rain** (Lovejoy and Mandelbrot 1985); **cracks** (Ali *et al* 1992, Xie 1995) and **fractal crack models** (Bouchaud *et al* 1993); **the structure of the universe** (Peebles 1989, Gurzadyan and Kocharyan 1991, Coleman and Pietronero 1992); **fractal surfaces** (Min *et al* 1995); **the ocean floor** (Herzfeld *et al* 1995); **the structure of cities** (Batty and Longley 1986, Batty 1995); **sunspots** (Milovanov and Zelenyi 1993); **cement structures and diffusion therein** (Niklasson 1993); the optical properties of **fractal quantum wells** (Gourley *et al* 1993); **growth forms of sponges** (Kaandorp 1991); **music** (Hsü and Hsü 1992); **oil and gas reserves** (Poon *et al* 1993); **medicine** (Keipes *et al* 1993); **biology and medicine** (Havlin *et al* 1995); **human retinal vessels** (Family *et al* 1989); the **classification of Chinese landscape drawings** (Voss 1992); **evolution** (Vandewalle and Ausloos 1995); **the shape of broccoli and cauliflower** (Grey and Kjems 1989); **microscopy** (Cross 1994); **food research** (Peleg 1993); **stereological measurements** (Flook 1982); **superconductors** (Wang *et al* 1994); **fungal morphology** (Crawford *et al* 1993); **ocean waves** (Zosimov and Naugol'nykh 1994); **turbulent fluid jets** (Flohr and Olivari 1994); **communications networks and the fractal structure of population distributions** (Appleby 1995); **turbulent flames** (Smallwood *et al* 1995); **graphite shapes in cast irons** (Lu and Hellawell 1994); **pressure transient analysis in naturally fractured reservoirs** (Acuna *et al* 1995, Acuna and Yortsos 1995); **fractal electrodes** (Bolz *et al* 1995); **quasi-brittle fracture** (Bazant 1995).

3.7.4 Perimeter–area relationship

The perimeter–area relationship has been used extensively to characterize many sets of fractal objects in a whole range of scientific and engineering problems. References which specifically cite the use of the perimeter–area relationship include those by Mandelbrot *et al* (1984) and Mu and Lung (1988), who characterize the contours of **fracture surfaces of steel**; Pande *et al* (1987), who characterize **fractured titanium alloy**; Issa and Hammad (1994), who similarly use the technique to characterize **concrete fractures**; Nikora *et al* (1993) and Wu and Lai (1994), who investigate **river channels and drainage areas**; Krummel *et al* (1987), who characterize **deciduous forest patterns**; Goodchild (1988), who investigates the similarity of **lake forms** on a simulated landscape; Lovejoy (1982), Hentschel and Procaccia (1984) and Rys and Waldvogel (1986), who characterize **rain clouds and hail clouds**. The derivation of the perimeter–area relationship is explained simply in Hastings and Sugihara (1993). Cheng (1995) gives an up to date overview of generalized perimeter–area relationships. This paper also contains a brief application of the technique to **geochemical problems**.

3.7.5 Erroneous dimension estimates

Finally, it should be mentioned that, under certain conditions, erroneous dimension estimates may be found for non-fractal data sets (Hamburger *et al* 1996) and great care should be taken when determining dimension estimates from real data.

3.8 Revision questions and further tasks

Q3.1 List the keywords and key phrases of this chapter and briefly jot down your understanding of them.

Q3.2 Use 3D hypercubes of side length δ to show that the box counting dimension of a plane lamina is two. It may help to draw a sketch using a square lamina with a side length of eight units, i.e. area = 64 units, and use values of δ which are powers of two, i.e. 2, 4, 8, etc. Plot $\log(N)$ against $\log(\delta)$ to obtain D_B.

Q3.3 (a) On graph paper show the construction of the triadic Cantor set to step $k = 4$. (Remember step $k = 0$ is the unit line.) Use 2D hypercubes of various side length δ to produce an estimate of the box counting dimension of the set. To do this plot $\log(N)$ against $\log(1/\delta)$ for various values of δ. Use values of δ which are integer powers of one third.

(b) Repeat (a) using a randomized triadic Cantor set generated by randomly selecting one third of the remaining line segments for removal at each stage in the construction (see the top of figure 3.1). How does the estimate compare with that obtained in (a)?

(c) What happens when you use δ's which are not integer powers of one third in (a) and (b) above?

Q3.4 Use the structured walk technique to show that the divider dimension of a regular triadic Koch curve is equal to the similarity dimension of 1.26.... Use the large Koch curve provided in figure Q3.4, normalizing all step lengths by the base length. First use step lengths which are integer powers of one third of the base length, then investigate for yourself the effect of using step lengths which are not integer powers of one third of the base length.

Q3.5 From the Richardson plot of figure 3.11 in the text, determine the dimension of both the Spanish–Portuguese border and the west coast of Britain. In addition, determine the divider dimension of a circle at small scales. Explain the answers you obtain.

Q3.6 Count the number of boxes (N) that intersect the coastline of Iceland in figure Q3.6. Do this for both the large and small boxes. From the results estimate the fractal dimension of the coastline.

Q3.7 Explain, with the aid of a diagram, what is meant by the structure and texture of a fractal curve.

Q3.8 Use the structured walk technique to estimate the **structural** and **textural** divider dimensions of the rugged boundary given in figure Q3.8. Use the inswing, outswing and alternate methods beginning at points A–C in turn (i.e. nine walks in total).

Q3.9 Use the structured walk technique to estimate the **structural** and **textural** divider dimensions of the two synthetic soot particles in figure Q3.9. Again, for each particle, use the inswing, outswing and alternate methods beginning at points A–C in turn.

Q3.10 Calculate the fractal dimension of the coastline of mainland Scotland (figure Q3.10) from Mallaig to Portpatrick.

Q3.11 Select other coastlines from an atlas and find both the box counting dimension and the divider dimension. See how the two estimates compare.

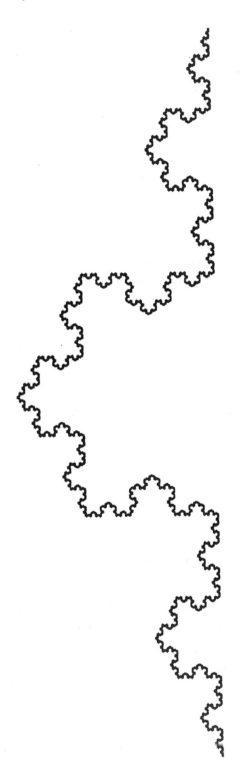

Figure Q3.4. The Koch curve.

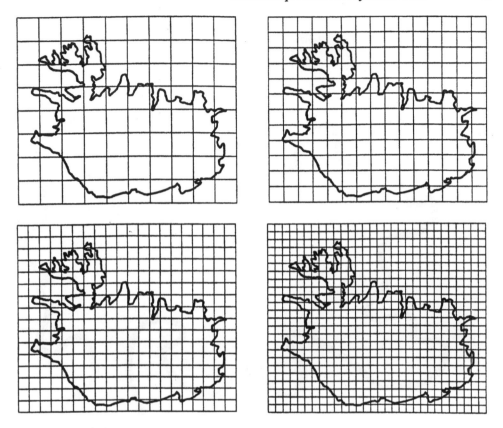

Figure Q3.6. The coastline of Iceland.

Q3.12 (*a*) Plot $\log(P)$ against $\log(A)$ for three squares of different side length. Using the perimeter–area relationship, find the dimension of the perimeter from the plot.

(*b*) Repeat (*a*) for equilateral triangles.

(*c*) Consider Koch islands (figure 2.14) with arbitrary initiator lengths. Verify that the perimeter–area relationship will produce a dimension $D_B = 1.2618\ldots$ for the coastline of such a group of Koch islands.

(*d*) Try to find a group of random fractal island shapes and see whether they are of the same shape (in a statistical sense) using the perimeter–area relationship.

Q3.13 Think of natural curves, other than those mentioned in the text, which you believe may have fractal properties. Try to list at least five.

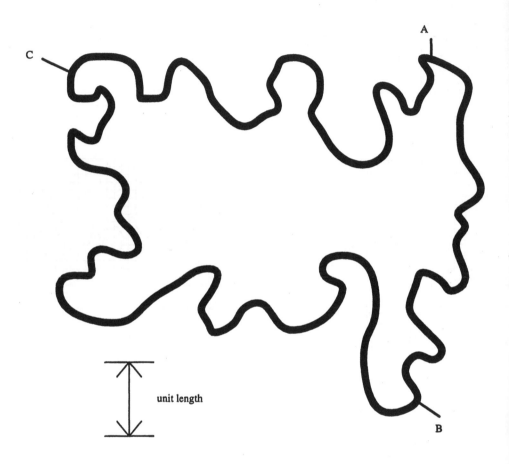

Figure Q3.8. A rugged boundary.

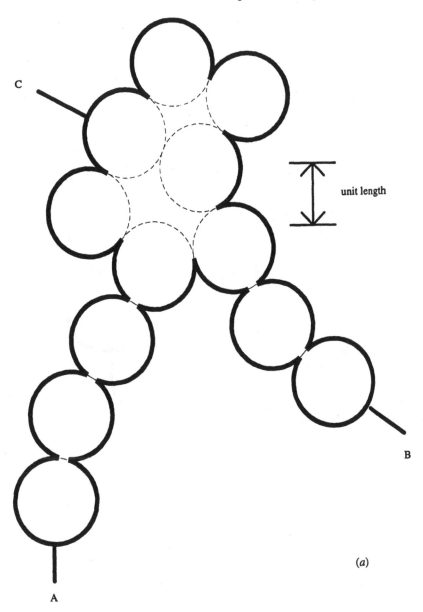

C

unit length

B

(a)

A

Figure Q3.9. (a) Synthetic soot particle 1. (b) Synthetic soot particle 2.

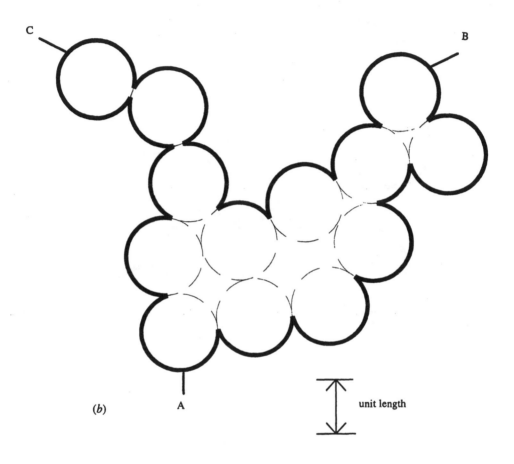

C

B

(b)   A   unit length

Figure Q3.9. *(b)* Continued.

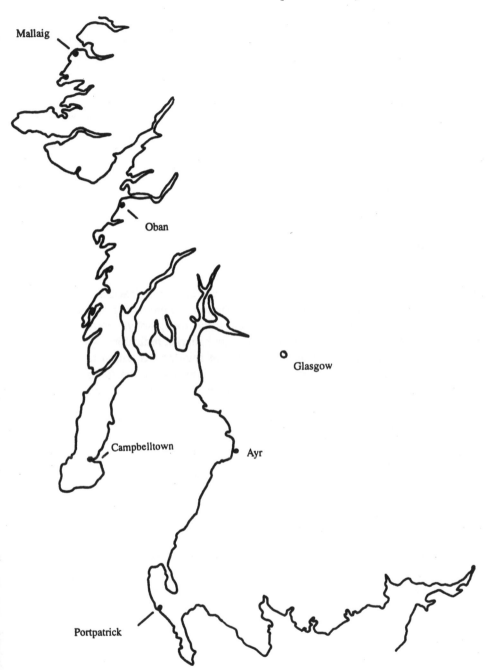

Figure Q3.10. The western coastline of Scotland.

Chapter 4

Regular and fractional Brownian motion

4.1 Introduction

In this chapter, we continue the discussion on random fractals begun in chapter 3. This time we concentrate on a specific set of random fractals known as **fractional Brownian motions** which have proved very useful in the description and modelling of many natural phenomena, including non-Fickian diffusion, landscape topography, DNA sequences, bacterial colonies, electrochemical deposition and stock market indicators. We will begin this chapter with the special (and well known) case of **regular Brownian motion**: then we move on to fractional Brownian motion proper and investigate the fractal properties of both its spatial **trajectories** and spatial–temporal (or spatio–temporal) **traces**.

4.2 Regular Brownian motion

Regular Brownian motion, or simply **Brownian motion**, is named in honour of the Scottish botanist Robert Brown (1773–1858) who, while using his microscope to observe pollen grains floating in water, noticed that they underwent rapid irregular motions. Brown found that other small particles also exhibited these seemingly unpredictable movements when placed on the water surface and he reasoned that the movement must be due to physical causes. We now know that the highly irregular motion of suspended particles at the water surface is due to their bombardment by the water molecules. Brownian motion is therefore a macroscopic manifestation of the molecular motion of the liquid. If we release a group of particles in a fluid at a specific location the action of the bombarding molecules in the liquid will cause the particles to spread out, or diffuse, through time. Molecular diffusion simulations based on Brownian motion are extensively used in science and engineering to model diffusion processes in both solid and fluid media.

An example of the trajectory of a particle undergoing Brownian motion in the plane is shown in the left-hand plot of figure 4.1(a). The particle was observed for 16 384 time steps and the beginning and end points of the particle's path are shown. Notice that, unlike the coastline fractals examined previously, Brownian motion trajectory curves in the plane may cross over themselves. As we zoom into the Brownian trajectory, more and more detail becomes apparent. This is illustrated by the right-hand plot of figure 4.1(a) which contains the first sixteenth of the left-hand plot magnified by an appropriate factor. The resolution of the two plots has been kept the same and the similarity between the two plots is evident. This statistical self-similarity extends over all scales of magnification. Brownian motion in the plane is in fact a random fractal curve.

If we only record the position of the Brownian motion at discrete time intervals, $\Delta t = t_1 - t_0$, $t_2 - t_1$, $t_i - t_{i-1}$, etc, and connect the observed points with straight line segments, we obtain a 'jerky' curve. This is shown in figure 4.1(*b*) where the sampled points are shown as dots. If we denote the positions of the sampled points at time t_i by (x_i, y_i) then the distribution of the observed jumps in the x and y co-ordinate directions, i.e. $\Delta x_i = (x_i - x_{i-1})$ and $\Delta y_i = (y_i - y_{i-1})$, follows a Gaussian (normal) probability distribution. In addition, the step lengths, $r_i \left(= \sqrt{(\Delta x_i)^2 + (\Delta y_i)^2} \right)$, between observed points, follow a Gaussian distribution. The two commonly used numerical methods to construct Brownian motion in the plane follow directly from these properties. We may randomly select steps in the two co-ordinate directions from a Gaussian distribution of known properties and build up the motion in the plane using Δx and Δy steps—the x, y method. Alternatively, we may randomly select the total step length, r, from a Gaussian distribution and randomly select the step angle, θ, from a constant distribution between 0 and 2π rad—the r, θ method.

The Brownian motion in figure 4.1(*a*) uses the x, y-method to generate the trajectory in the plane. The left-hand plot of figure 4.1(*a*) contains Brownian motion made up of 16 384 random steps taken in both the x and y co-ordinate directions (i.e. 32 768 steps in all): every sixteenth step in the x and y directions is plotted. The steps are taken from the Gaussian probability distribution of zero mean and unit variance shown in figure 4.1(*d*). The right-hand plot of figure 4.1(*a*) contains the first 1024 points of the left-hand trajectory: this time every step is plotted, giving the same resolution for both plots. As mentioned above, the similarity of the motion at the two scales is visually apparent, and in fact, if the scales were left off the plots it would be impossible to tell which plot was the original and which was the zoomed section due to the statistical self-similarity of the fractal trajectory curve.

Figure 4.1(*c*) uses the r, θ method to generate Brownian motion. This time 16 384 steps are taken from the Gaussian distribution of figure 4.1(*d*) and a random angle chosen for each step as described above. Notice that the trajectory of figure 4.1(*a*) seems to require a larger area to contain its wanderings than that of figure 4.1(*c*). This is because we used twice as many steps in the former figure and the expected displacement from the origin of a Brownian trajectory scales with the square root of the number of steps taken. We will return to the scaling properties of Brownian motion later in this section. Both Brownian motion plots of figures 4.1(*a*) and (*c*) take steps from the Gaussian distribution of zero mean and unit standard deviation shown in figure 4.1(*d*). We will use this distribution in all subsequent examples of Brownian and fractional Brownian motion in this chapter.

The term Brownian motion has become ambiguous as it has been used to describe both spatial-only curves, such as the plots of Brownian motion in the plane of figures 4.1(*a*) and (*c*), and spatial–temporal plots, e.g. time histories of the components of Brownian motion. Herein, to avoid confusion, we will refer to these curves as **trajectories** in the spatial-only case and **traces** in the spatial–temporal case. (Note that

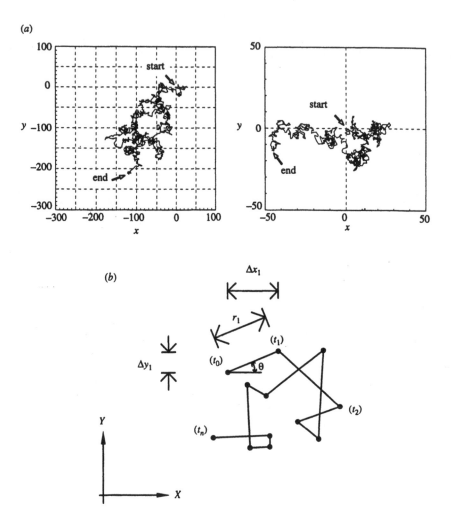

Figure 4.1. Properties of Brownian motion in the plane. (*a*) Brownian motion modelled using Brownian increments in x and y co-ordinate directions (left, $0 < t < 16\,384$ sampled at a time interval $\Delta t = 16$; right, $0 < t < 1024$ sampled at $\Delta t = 1$. (After Addison and Bo (1996). Reprinted from Chen Ching-Jen, Shih C, Lienau J and Kung R J (eds) 1996 *Flow Modeling and Turbulence Measurements VI—Proc. 6th Int. Symp. (Tallahassee, FL, 1996)* 932 pp, Hfl. 230/US$135/£94.00. A A Balkema, PO Box 1675, Rotterdam, The Netherlands. (*b*) Brownian motion sampled at a regular time interval where $r_i = \sqrt{(\Delta x_i)^2 + (\Delta y_i)^2}$. (*c*) Brownian motion in the x–y plane (left, $0 < t < 16\,384$ sampled at $\Delta t = 16$; right, $0 < t < 1024$ sampled at $\Delta t = 1$. (*d*) The distribution of the step lengths used in both the x, y method (for the x and y direction steps) and the r, θ method (for the r steps).

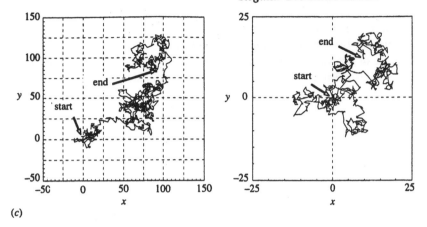

(c)

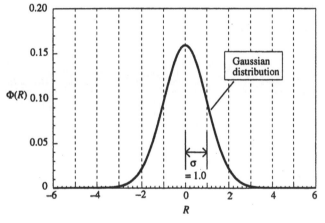

(d)

Normal, or Gaussian, probability

density function (p.d.f.) of the form: $\Phi(R) = \dfrac{1}{\sqrt{2\pi}}\ e^{-\frac{1}{2}(R^2)}$

The probability of the step length occurring in the range $R_a \leqslant R \leqslant R_b$ is:
$P(R_a \leqslant R \leqslant R_b) = \int_{R_a}^{R_b} \Phi(R)\,dR$

Standard deviation, $\sigma = 1.0$; zero mean

Figure 4.1. Continued.

these Brownian curves may also be found elsewhere in the literature under the names of Brownian trails and Brownian functions respectively.)

A time trace of Brownian motion in one co-ordinate direction, $B(t)$, is given in figure 4.2(c). This trace is the time history of one of the co-ordinates of a Brownian trajectory. We saw above that successive increments of the Brownian trace, $B(t) - B(t - \Delta t)$, have a Gaussian distribution. From this observation we see that a discretized approximation to a Brownian motion trace, $B(t_i)$, may be produced at discrete times, $t_i = i \Delta t$ (where i is an integer), by summing up a series of random steps, $R(t_j)$, taken from a Gaussian distribution (such as the one given in figure 4.1(d)) as follows:

$$B(t_i) = \sum_{j=1}^{i} R(t_j) \tag{4.1}$$

The construction process of the Brownian trace in figure 4.2(c) is illustrated in figures

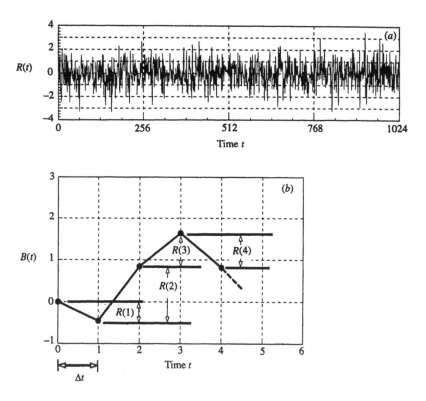

Figure 4.2. The construction and properties of the time trace of Brownian motion. (a) 1024 discrete random increments taken from a Gaussian distribution with unit variance sampled at unit incremental time. (b) The construction process of a Brownian motion trace. (c) Brownian motion trace. (d) Zooming into the middle quarter of the Brownian motion trace in (c) above (zoom 1). (e) Zooming in again—zoom 2 in (c) above. (Continued overleaf).

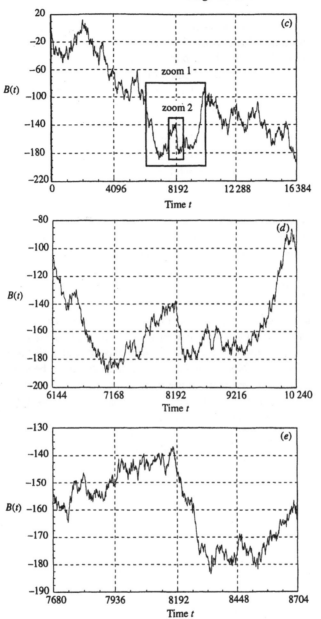

Figure 4.2. Continued.

4.2(a) and (b). Figure 4.2(a) shows the first 1024 discretely sampled random numbers used in the construction of the Brownian trace. These numbers are taken from the Gaussian distribution of figure 4.1(d), and are used as the incremental steps, $R(t_j)$, in the Brownian motion approximation of equation (4.1). Figure 4.2(b) shows the first four steps of the Brownian trace generated using the first four Gaussian random numbers of figure 4.2(a). At the high resolution of figure 4.2(b) the discrete steps are obvious (cf figure 4.1(b)). However, if we zoom out far enough, we lose sight of the fine detail required to resolve the individual steps and are left with a prefractal curve (figure 4.2(c)) which is indistinguishable, from the viewer's stand point, from a continuous Brownian motion time trace—a true fractal, possessing self-similar detail at all levels of magnification. The finite resolution of the generated Brownian motion trace is therefore governed by the choice of the time increment Δt. This is similar to the production of the fractal curve illustrations in chapter 2, where we undertook as many steps as required in the construction process to produce enough detail to 'fool the eye', since it is not possible to reproduce the infinite detail of a true fractal object.

Now we have generated a prefractal Brownian trace (figure 4.2(c)) let us, for arguments sake, assume it is an actual Brownian trace with infinite detail, i.e. $B(t)$ where the time, t, is continuous. The self-similarity of $B(t)$ is evident as we zoom into it. Zooming into the middle one quarter section of figure 4.2(c) we arrive at figure 4.2(d). Further zooming in to the middle one quarter section of figure 4.2(d) we arrive at figure 4.2(e). The original and zoomed traces look similarly irregular to the eye, and in fact are statistically self-similar. Notice, however, that the axes have been scaled differently at each magnification. The segment of the time trace under observation was scaled up four times, however, the spatial co-ordinate, $B(t)$, was scaled up by only a factor of two. Thus, two different scaling factors are required to retain the self-similarity of the original trace and the scaled trace. This anisotropic scaling requirement for self-similarity in Brownian traces is considered in more detail below.

If we consider pairs of points on a trace of Brownian motion separated by a

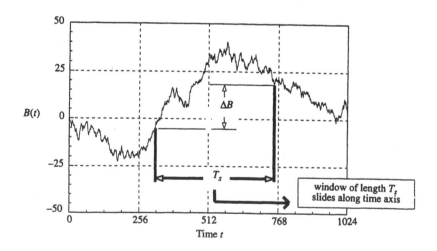

Figure 4.3. The scaling of increments on a Brownian motion trace.

time T_S, we find that the mean, absolute separation in $B(t)$ between points, i.e. $\overline{|\Delta B|} = \overline{B(t + T_S) - B(t)}$, is proportional to the square root of the time of separation, i.e.

$$\overline{|\Delta B|} \propto T_S^{1/2} \qquad (4.2)$$

where the exponent, equal to one-half, is known as the **Hurst exponent**, H. The average value of the mean separation is found by sliding a time window of length T_S over a sufficiently large portion of the trace and averaging $|\Delta B|$: figure 4.3 illustrates the method. For Brownian motion, the sum of the statistically independent spatial steps in time leads to a mean divergence of the separated points which scales with the square root of time. A Brownian motion trace, therefore, remains statistically self-similar under scalings only when the axes $B(t)$ and t are magnified by different factors. That is, if we scale t by a factor A, we must scale $B(t)$ by a factor A^H. This property of non-uniform, or anisotropic, scaling is known as **self-affinity**. Self-affinity is the reason for the two scaling factors involved when plotting the magnified sections of the figure 4.2(c) time series in figures 4.2(d) and (e). When we zoomed into one quarter of the time series of figure 4.2(c) to produce figure 4.2(d), the time axis was magnified by a factor of four, but the $B(t)$ axis was only scaled up by a factor of $(4)^{1/2} = 2$. This retained the statistical self-similarity of the time trace between the two figures.

Figure 4.4(a) illustrates the diffusion through time of particles undergoing Brownian motions. For clarity, only ten traces are plotted. At $t = 0$ all ten traces start at $B(t) = 0$ and begin noticeably to spread out from the origin as time increases. This spreading is in an average sense, as it can be seen that some of the traces do return to $B(t) = 0$: some traces even return a few times. If, instead of ten Brownian traces, we use a far larger number, then we can characterize the spreading process using averaged statistical properties. Imagine a much larger number of released particles diffusing away from the origin through time in figure 4.4(a). When considering diffusion related problems, it is more appropriate to deal with standard deviation as a length measure of the resulting diffusion cloud. The standard deviation σ_C is proportional to $\overline{|\Delta B|}$ and the constant of proportionality depends upon the probability distribution of the random function $R(t)$. We find that the standard deviation of the diffusing cloud, σ_C, scales in the same manner as $\overline{|\Delta B|}$, that is

$$\sigma_C \propto t^{1/2}. \qquad (4.3a)$$

The variation of σ_C with time for a diffusion cloud of many particles is shown schematically in figure 4.4(b) (it is also superimposed on figure 4.4(a) plotted for the many particle case). Diffusive processes which scale in this manner are known as **Fickian**. For example, molecular diffusion of a dye in a still liquid is a Fickian process. It is usual to express the scaling relationship as the equality

$$\sigma_C = \sqrt{2Kt} \qquad (4.3b)$$

where K is known as the **diffusion coefficient**, with units of square metres per second. In practice, the diffusion coefficient is obtained from diffusing clouds or plumes, by plotting the variance, σ_C^2, against time (see figure 4.4(c)). This results in a linear relationship and the best fit line has a slope equal to twice the diffusion coefficient.

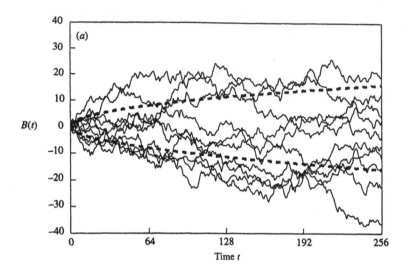

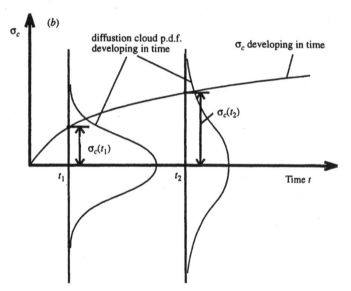

Figure 4.4. The scaling of a diffusing cloud of Brownian motion particles. (*a*) The traces of ten Brownian motions all starting at the origin. (The development of the expected standard deviation with time, resulting from a particle cloud with a very large number of particles, is plotted dashed either side of the origin.) (*b*) The scaling of a cloud of many particles with time. (*c*) A variance against time plot showing the best fit line used to obtain the diffusion coefficient from experimental data (shown as dots) for Fickian diffusion. (Continued overleaf).

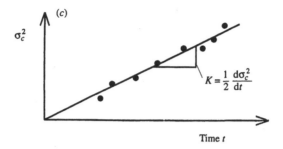

Figure 4.4. Continued.

We can generate $B(t)$ with a specific diffusion coefficient K by taking the incremental steps, $R(t_j)$, from a Gaussian distribution with standard deviation, σ_p, given by

$$\sigma_p = \sqrt{2K\Delta t} \tag{4.3c}$$

then after i time steps, each Δt in duration, it follows from equations (4.3b, c) that the standard deviation of a diffusing cloud of particles will be

$$\sigma_c = \sigma_p(i)^{1/2} \tag{4.3d}$$

Coarser approximations to Brownian motion are often produced using simpler probability distributions for $R(t_j)$. Two popular methods use either a constant probability density function (p.d.f.), or simple discrete steps: the latter simply 'walks off' using a step length of $\pm L$ chosen at random. The standard deviation of the resulting particle cloud generated from these simpler probability distributions follows the scaling given by equations (4.3b–d). However, the shape of the resulting particle cloud changes from the original (simple) probability distribution, tending to a Gaussian distribution as time increases: this is a manifestation of the central limit theorem (refer to any good statistics book for more information). In addition, individual time traces of the motion appear more and more like Brownian motion as one zooms out, again due to the central limit theorem. Figure 4.5 illustrates the construction of a random walk using these simpler probability distribution functions, both having unit standard deviation and zero mean. In this figure, both the first 64 and the first 1024 steps are plotted for each distribution. We can see that the traces appear more Brownian-like as the number of steps increases. The Brownian motion constructed using the $\pm L$ (delta) distribution is sometimes referred to as a **random walk**, staggered walk or drunken man's stagger due to the nature of the random steps taken through time. As a Brownian time trace may be thought of as the time history of one co-ordinate of a Brownian trajectory, then to construct a trajectory in the plane (respectively 3D space) we simply use equation (4.1) two (respectively three) times, generating each co-ordinate independently. We will implement this in detail for the more general case of fractional Brownian motion in section 4.5.

4.3 Fractional Brownian motion: time traces

Brownian motion is in fact a special member of a larger family known as **fractional Brownian motions,** usually abbreviated to fBms, a generalization of Brownian motion

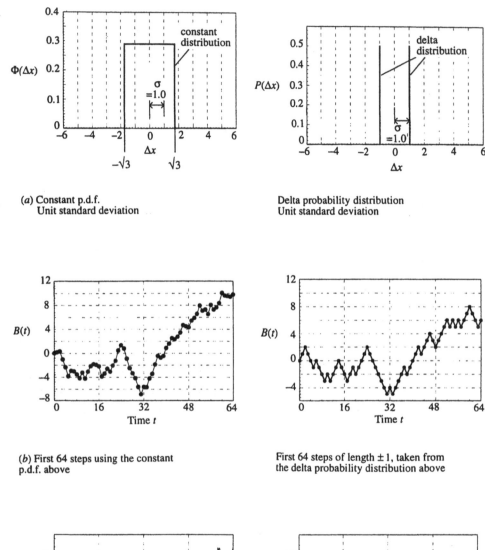

(a) Constant p.d.f.
Unit standard deviation

Delta probability distribution
Unit standard deviation

(b) First 64 steps using the constant
p.d.f. above

First 64 steps of length ± 1, taken from
the delta probability distribution above

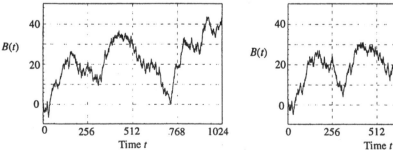

(c) First 1024 steps (constant p.d.f.)

First 1024 steps (delta distribution)

Figure 4.5. Random walks generated from constant and delta probability functions.

suggested by Mandelbrot. fBms have Hurst exponents in the range $0 < H < 1$, where the special case of $H = 0.5$ gives regular Brownian motion as discussed above. As mentioned at the beginning of this chapter the concept of fBm has gone a long way to assist in the understanding and modelling of irregular phenomena in science and nature, e.g. mountainous terrains, structural cracking (see hillside, skyline and wall cracks of figure 1.2), stockmarket indicators, particle paths in fluids, etc. As with Brownian motion, we differentiate between spatial–temporal traces and solely spatial trajectories of the motion. We will investigate fBm traces and trajectories separately, dealing with fBm traces first and considering fBm trajectories in section 4.5.

We shall denote fBm by $B_H(t)$ where the subscript H is the Hurst exponent which classifies the motion. fBm traces for $H = 0.2$, 0.5 and 0.8 are shown in figure 4.6 (recall that $H = 0.5$ corresponds to regular Brownian motion). As can be seen from the figure, there is a noticeable, qualitative difference between each trace. Traces corresponding to values of H less than 0.5 have a tendency to turn back upon themselves: this property is known as **antipersistence**. On the other hand, for values of H greater than 0.5, the trace has a tendency to persist in its progression in the direction in which it was moving: this behaviour is known as **persistence**.

In a similar fashion to the regular Brownian motion discussed in the previous section, fBms are self-affine processes. If we scale up a portion of an fBm curve, then to preserve statistical self-similarity we require two independent scaling factors for the t and $B_H(t)$ axes. Scaling t by a factor of A requires that $B_H(t)$ must be scaled by A^H, thus t becomes At and $B_H(t)$ becomes $A^H B_H(t)$. The mean absolute separation along an fBm trace, $\overline{|\Delta B_H|}$, scales with time of separation on the trace as

$$\overline{|\Delta B_H|} \propto T_S^H \tag{4.4}$$

and similarly the standard deviation of a diffusing cloud of fBm particles scales as

$$\sigma_c \propto t^H. \tag{4.5a}$$

Expressions (4.4) and (4.5a) are basically expressions (4.2) and (4.3a), where the Hurst exponent of one-half, corresponding to regular Brownian motion, has been replaced with the more general H of fBm, where $0 < H < 1$. Following on from expression (4.5a), a **fractal diffusion coefficient** K_f may then be found by plotting $(\sigma_c)^{\frac{1}{H}}$ against time since release, t, and defining

$$K_f = \frac{(\sigma_c)^{\frac{1}{H}}}{2t}. \tag{4.5b}$$

This time, however, the units of K_f depend upon H. fBm particles exhibit **non-Fickian (or anomalous) diffusion** when $H \neq \frac{1}{2}$. A plot of $(\sigma_c)^{\frac{1}{H}}$ against time will produce the linear relationship shown in figure 4.7(a) where the slope of the line is equal to twice the fractal diffusion coefficient. In practice, both H and K_f may not be known. If this is the case, then the best fit line through the experimental data points drawn on a logarithmic plot of σ_c against time may be used (see figure 4.7(b)). The gradient of the best fit line is H and its crossing point on the σ_C axis is $H \log(2K_f)$.

If an fBm is a statistically self-similar process with structure at all scales the natural question to ask is 'What is its fractal dimension?'. Here we shall consider the fBm time

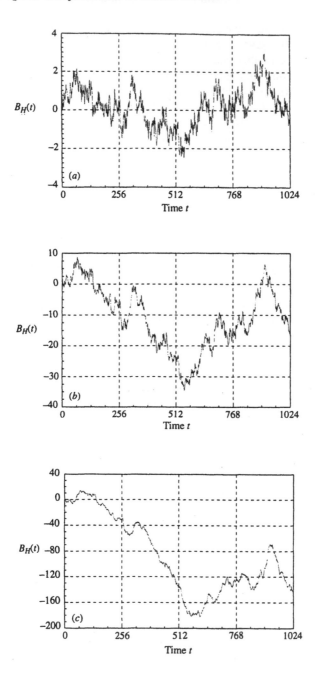

Figure 4.6. Fractional Brownian motion time traces (various Hurst exponents). (*a*) fBm trace, $H = 0.2$—antipersistent. (*b*) fBm trace, $H = 0.5$—neutrally persistent (regular Brownian motion). (*c*) fBm trace, $H = 0.8$—persistent.

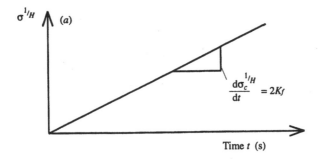

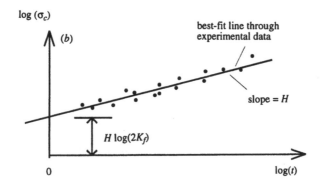

Figure 4.7. The fractal diffusion coefficient for fBm diffusion. (After Addison (1996). Reproduced by permission of the IAHR). (*a*) The relationship between standard deviation and time for fBm. (*b*) The experimental derivation of H and K_f.

trace and leave a discussion of the fractal dimension of the spatial trajectory to section 4.5. The box counting fractal dimension, D_B, of an fBm trace may be obtained by considering a portion of an fBm trace which has been rescaled for simplicity so that it fits into a unit box, i.e. $0 \leq t \leq 1$ and $0 \leq B_H(t) \leq 1$, see figure 4.8. We then divide the time interval into n segments each of length $1/n$. The height of the box required to contain the time trace over the time sub-interval $1/n$ will scale differently to its base due to the self-affine nature of the trace. Referring back to the above discussion concerning the self-affinity of the time trace we see that the scale factor A corresponds to $1/n$, hence the box height must scale from the original range of $B_H(t) = 1$ to $(1/n)^H \times 1 = (1/n)^H$. The area of each of these rectangular boxes, required to cover the trace at each sub-interval of time, is simply the length multiplied by the breadth of the box, or

$$\left(\frac{1}{n}\right)\left(\frac{1}{n^H}\right) = \frac{1}{n^{H+1}}. \tag{4.6}$$

We need, however, to relate the scaling of smaller square boxes to the original square box of unit side length. Thus, the typical number of smaller squares, of area $= (1/n)^2$,

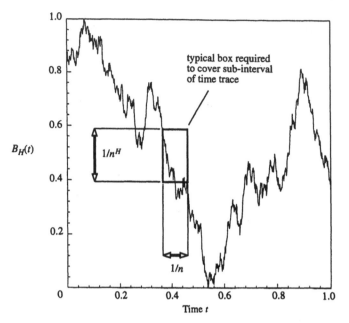

Figure 4.8. Scaling properties of an fBm trace contained within the unit box (persistent fBm—$H = 0.6$).

required to cover the trace in each time sub-interval is

$$\left[\left(\frac{1}{n^{H+1}}\right) \middle/ \left(\frac{1}{n^2}\right)\right] = \frac{1}{n^{H-1}}. \tag{4.7}$$

There are n such sub-intervals of time each $(1/n)$ in length. Therefore, the number of boxes, N, of side length $\delta = 1/n$ required to cover the original portion of the fBm in the unit box is

$$N = n\left(\frac{1}{n^{H-1}}\right) = \frac{1}{n^{H-2}}. \tag{4.8}$$

Relating this to the definition of the box counting dimension given in equation (3.1), we obtain

$$D_B = \frac{\log(N)}{\log(1/\delta)} = \frac{\log\left(\frac{1}{n^{H-2}}\right)}{\log(n)} = \frac{(2-H)\log(n)}{\log(n)}. \tag{4.9a}$$

Cancelling out the $\log(n)$ terms leaves the following simple expression for the box counting dimension of an fBm trace:

$$D_B = 2 - H. \tag{4.9b}$$

Thus, the fractal dimensions of the Brownian motion traces of figure 4.6, given by the box counting dimension, are the following: for $H = 0.2$, $D_B = 1.8$; for $H = 0.5$ (regular

Brownian motion), $D_B = 1.5$; and for $H = 0.8$, $D_B = 1.2$; i.e. the more convoluted the trace the higher the dimension. This seems to fit, in some way, with our intuitive understanding of the dimension estimate being a measure of the 'crinklyness' or 'degree of convolution' of a curve.

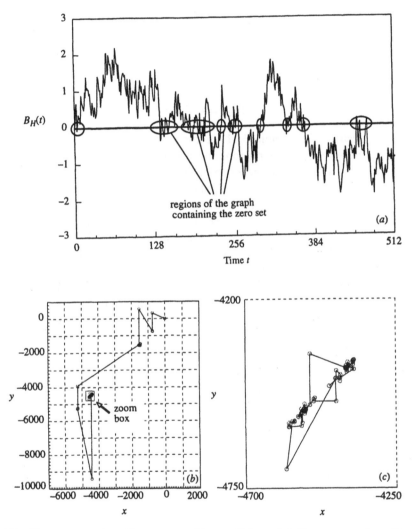

Figure 4.9. The zeroset of an fBm trace and Levy flights. (*a*) The zeroset of an fBm trace ($H = 0.2$, i.e. antipersistent). (*b*) Levy flight ($D_B = 0.5$). (*c*) Magnified box of (*b*).

If we slice through an fBm curve at $B_H(t) = 0$ we obtain a set of points known as the **zeroset** of the fBm. The zeroset is then the set of all the points where the fBm curve crosses the time axis (figure 4.9(*a*)). (In fact we can generate many similar sets, known as isosets, by slicing through the fBm at any constant value of $B_H(t)$.) The zeroset is a set of disconnected points with a (random) Cantor-set-like construction. By sampling only those points in the trace which cut the time axis we essentially remove a spatial dimension from the fractal, thus the box counting fractal dimension of the set of crossing points becomes

$$D_B = 1 - H. \tag{4.10}$$

As one can see from figure 4.9(a), the fBm trace (here $H = 0.2$) crosses the time axis intermittently. Where the trace does cross the time axis it is likely to do so again many more times over a short time interval. Thus, the zeroset is composed of clusters. The probability of observing a time interval between two subsequent crossings, T_c, has a cumulative power law distribution of the form

$$P(T_c) = CT_c^{-D_B} \tag{4.11}$$

where C is a constant. Expression (4.11) is known as the Levy distribution. One may use Levy distributions to generate random power law trajectories known as **Levy flights**. Figure 4.9(b) shows a Levy flight generated from the zerosets of an fBm with $H = 0.5$ (hence $D_B = 0.5$). The trajectory (or flight) is generated in a similar manner to the Brownian motion trajectory in figure 4.1(c), this time however, the random steps are taken from the zerosets of an fBm with $D_B = 0.5$, i.e. a Levy distribution. In figure 4.9(b), the end point of each random step is marked by a small circle. We see from the figure that large step lengths are relatively rare, however, they do dominate the resulting trajectory. By zooming into clusters of points (e.g. figure 4.9(c)), we may see statistically self-similar structure at all scales. It has been noted that the clustering of the end points in Levy flights with exponents D of 1.26 is very similar to the distribution of galaxies in the universe when observed from earth.

One remaining problem is: 'How do we actually generate an fBm time series?' An fBm trace is defined at time t as

$$B_H(t) = \frac{1}{\Gamma\left(H + \frac{1}{2}\right)} \left\{ \int_{-\infty}^{0} \left[(t - t')^{H - \frac{1}{2}} - (-t')^{H - \frac{1}{2}} \right] dB(t') \right.$$
$$\left. + \int_{0}^{t} (t - t')^{H - \frac{1}{2}} dB(t') \right\} \tag{4.12}$$

where $B(t)$ is continuous regular Brownian motion and Γ is the gamma function. In fact, the generation of a discrete fBm trace is a much more involved process than that for ordinary Brownian motion. We cannot simply add a single random number at each stage in the construction of an fBm as each step is not independent, but rather depends upon the whole history of the fBm to that point. Or, to put it another way, an fBm has an infinitely long memory associated with it. We can generate an approximation to an fBm using successive random increments from a Gaussian distribution and a finite memory. The following two steps generate a quick and reasonably good approximation to an fBm (especially with $H > 0.5$).

(i)   Each incremental step in the construction process is generated using

$$B_H(t_i) - B_H(t_{i-1}) = \frac{1}{\Gamma\left(H + \frac{1}{2}\right)} \left[\sum_{j=i-M}^{i-2} \left((i \quad j)^{H - \frac{1}{3}} \quad (i - j - 1)^{H - \frac{1}{2}} \right) R(t_j) \right.$$

$$+ \sum_{j=i-1}^{i} (i-j)^{H-\frac{1}{2}} R(t_j) \Bigg]. \qquad (4.13a)$$

(ii) The fBm at time t_i is found from a summation of these incremental steps, as follows:

$$B_H(t_i) = \sum_{k=1}^{i} \big[B_H(t_k) - B_H(t_{k-1}) \big]. \qquad (4.13b)$$

$B_H(t_i)$ is the ith discrete approximation to the fBm at time $t_i = \Delta t \, i$; Δt is the discrete time step used; H is the Hurst exponent of the trace; M is the number of steps of the limited memory used in the approximation of the fBm; and $R(t_i)$ are random steps sampled discretely from a Gaussian probability distribution (e.g. figure 4.1(d)). Equation (4.13a) has been written in a standard form and the second term within the square brackets simply reduces to $R(t_{i-1})$. Note that there are many alternative methods for the generation of fBms and the sophistication of the technique used depends upon the accuracy required. See the further reading section at the end of this chapter for more details.

If σ_p is the standard deviation of $R(t_i)$, then after i steps, the standard deviation of a cloud of fBms σ_c is given by

$$\sigma_c = \sigma_p (i)^H \qquad (4.14a)$$

which for a Hurst exponent of 0.5 reduces to equation (4.3c) for regular Brownian motion. Given a fractal diffusion coefficient, we may rearrange equation (4.5b), to define σ_p in our fBm approximation as

$$\sigma_p = \big(2K_f \, \Delta t \big)^H. \qquad (4.14b)$$

To put the fBm generation method of equations (4.13a, b) in practical terms: if we want to generate N steps of an fBm, $B_H(t)$, then we require $N + M$ random steps from a Gaussian distribution, i.e. we require $R(t_i)$ where i is from zero to $N + M$. To obtain a reasonable approximation to the fBm requires M to be at least as large as N. (The larger the M/N ratio, the better.) There are two things to note about expressions (4.13a, b): firstly the fBm generated using them passes through the origin, i.e. $B_H(t_0) = 0$, and secondly it reduces to regular Brownian motion (4.1) when $H = 0.5$. The construction process is shown in figure 4.10, where 512 points of an fBm, with $H = 0.8$, are generated using 1024 random numbers, $R(t)$, taken from a Gaussian distribution, with zero mean and unit standard deviation. The generation of the ith point in the fBm trace requires the summation of all the random numbers over the time interval from t_{i-M} to t_i, corresponding to a temporal memory of $M\Delta t$. This is shown in figure 4.10, where, for reasons of clarity in the illustration, we have used $M = N$, i.e. 512. (Remember $M \gg N$ is better.)

4.4 Fractional Brownian surfaces

fBm trace functions may be extended to higher dimensions where the fBm is a function of more than one t variable, i.e. $B_H(t_1, t_2, t_3, \ldots, t_N)$. Here we should think of the t_i's

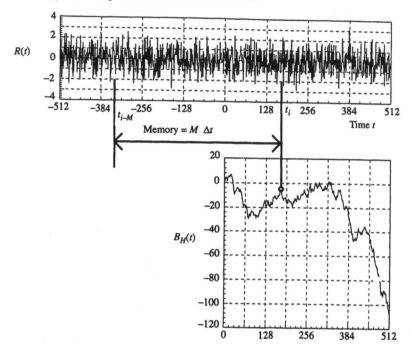

Figure 4.10. Graphical illustration of the construction of an fBm trace. All the random numbers within the memory window are used to generate the circled point on the fBm trace.

as the independent variables upon which the multi-dimensional fBm depends, and not times. This generalization of the trace function has the following properties:

(i)   The increments $\Delta B_H = B_H(t_1, t_2, t_3, \ldots, t_N) - B_H(t_1', t_2', t_3', \ldots, t_N')$ are Gaussian with zero mean.

(ii) The standard deviation of the increments, σ_B, is directly related to the separation between points, T_s, through the Hurst exponent as follows:

$$\sigma_B \propto T_s^H \tag{4.15a}$$

where, for the multi-dimensional fBm, the separation between points is given by

$$T_s = \sqrt{\sum_{i=1}^{N} \left(t_i - t_i'\right)^2}. \tag{4.15b}$$

Trace functions of the form $B_H(t_1, t_2)$ have been successfully employed in the generation of fractal models of natural landscapes and internal crack surfaces. Figure 4.11 contains 3D plots of $B_H(t_1, t_2)$ trace functions for various Hurst exponents. In this figure, the vertical component, $B_H(t_1, t_2)$, is plotted against t_1 and t_2 in the horizontal plane. The height of the simulated landscapes is then $B_H(t_1, t_2)$ and is a function of its horizontal location given by the spatial t variables. Figures 4.11(a–c) contain Brownian landscapes with Hurst exponents of 0.2, 0.5 and 0.8 respectively. The landscape surface becomes

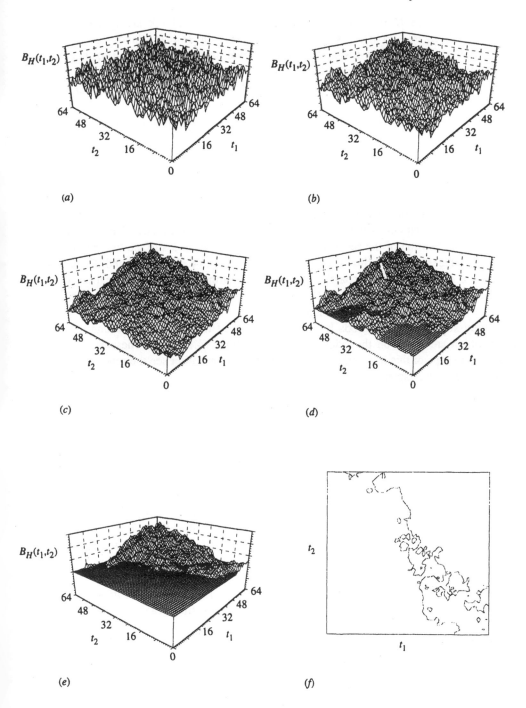

Figure 4.11. Fractional Brownian surfaces. (*a*) Brownian surface $H = 0.2$. (*b*) Brownian surface $H = 0.5$. (*c*) Brownian surface $H = 0.8$. (*d*) Brownian surface $H = 0.8$ (with 'sea' at low tide). (*e*) Brownian surface $H = 0.8$ (high tide). (*f*) Coastline contour of plot (*e*).

visibly smoother as H increases. Hurst exponents of around 0.8 are typical of real landscapes. In figure 4.11(d), a 'sea' has been added to the $H = 0.8$ landscape in figure 4.11(c), i.e. all points below a specified surface level have been set equal to this level. The sea level has been increased in figure 4.11(e). Figure 4.11(f) contains a plot of the coastline (i.e. the zero-contour) of the landscape of figure 4.11(e). Figure 4.11(f) is in fact the zeroset of $B_H(t_1, t_2)$.

The fractal dimension of fBm trace functions of N variables is given simply by

$$D_B = N + 1 - H \qquad (4.16)$$

where $0 < H < 1$. (This reduces to equation (4.9b) for the fBm time traces dealt with in the last section where $N = 1$.) Thus, the modelled Brownian landscape of figure 4.11(c) has a fractal dimension of 2.2. In addition, the coastline plot of figure 4.11(f) is in effect a horizontal cut of the surface (at the water level) by a plane. This reduces the topological dimension of the resulting object by one (i.e. from a surface to a curve), and similarly reduces its fractal dimension by one (cf equation (4.10) and related discussion). The fractal dimension of the coastline is then 1.2. Furthermore, any vertical cut of the surface results in a single variable fBm trace function again, just like those considered in section 4.3, with a fractal dimension of 1.2.

Many sophisticated methods have been developed to generate Brownian surfaces, all of which are based on the rules given by equations (4.15a, b). These methods are outside the scope of this introductory text. For more information see the references given at the end of this chapter.

4.5   Fractional Brownian motion: spatial trajectories

In section 4.3, we considered spatial–temporal fBm traces, i.e. $B_H(t)$ against t (and we extended this in section 4.4 to $B_H(t_1, t_2, t_3 \ldots, t_N)$). However, $B_H(t)$ is only the time history of a 1D process. An fBm trajectory in the plane may be constructed using two independent fBm traces in the same way that we produced regular Brownian motion in figure 4.1(a) using the x, y method. The process is illustrated in figure 4.12(a) where the spatial components of two fBm traces, $H = 0.2$, are used as the x and y components of the fBm trajectory in the plane. Figures 4.12(b) and (c) contain fBm trajectories constructed using the same method with Hurst exponents of 0.5 and 0.8 respectively.

In contrast to fBm traces, where the curve is continuously marched through time with random displacements added at each time interval, fBm trajectories may cross over themselves as they wander about the plane. As with fBm time traces, fBm trajectories are also random fractals with statistical self-similarity at all scales. However, the fractal dimension for an fBm trajectory is different from the corresponding trace function. The box counting fractal dimension, D_B, of an fBm trace may be obtained by considering an fBm trajectory observed over a time interval T. This is shown schematically in figure 4.13. Referring back to equation (4.4), each co-ordinate of an fBm trajectory will vary typically over time sub-intervals Δt, by $\overline{|\Delta B_H|} \propto \Delta t^H$. On a trajectory, the number of boxes, N, of side length, δ, required to cover the trajectory scales simply with the 'size' or 'amount' of the trajectory which in turn is related to the time span of the motion, T, i.e. $N \propto T/\Delta t$. (Remember the trajectory does not mix temporal and spatial scales, i.e.

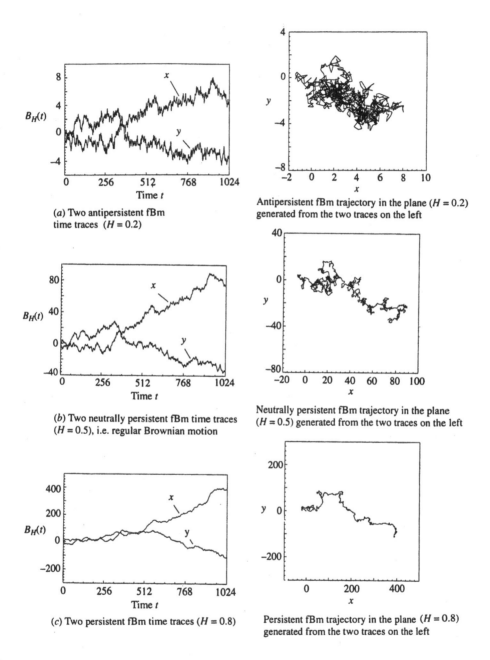

(a) Two antipersistent fBm time traces $(H = 0.2)$

Antipersistent fBm trajectory in the plane $(H = 0.2)$ generated from the two traces on the left

(b) Two neutrally persistent fBm time traces $(H = 0.5)$, i.e. regular Brownian motion

Neutrally persistent fBm trajectory in the plane $(H = 0.5)$ generated from the two traces on the left

(c) Two persistent fBm time traces $(H = 0.8)$

Persistent fBm trajectory in the plane $(H = 0.8)$ generated from the two traces on the left

Figure 4.12. The generation of two component fBms in the plane. $((c)$ After Addison and Bo (1996). Reprinted from Chen Ching-Jen, Shih C, Lienau J and Kung R J (eds) 1996 *Flow Modeling and Turbulence Measurements VI—Proc. 6th Int. Symp. (Tallahassee, FL, 1996)* 932 pp, Hfl. 230/US\$135/£94.00. A A Balkema, PO Box 1675, Rotterdam, The Netherlands.)

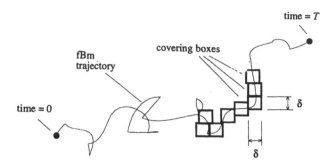

Figure 4.13. Scaling properties of an fBm trajectory.

it is not self-affine but rather self-similar, consisting of two spatial axes, both of which have the same scaling properties.) The spatial dimension of each box required to cover the fBm trajectory δ ($= \overline{|\Delta B_H|}$) scales with Δt^H. Thus, as we reduce the side length, δ, of the probing boxes, we obtain for the box counting dimension (choosing suitable units),

$$D_B = \frac{\log(N)}{\log(1/\delta)} = \frac{\log(1/\Delta t)}{H \log(1/\Delta t)} = \frac{1}{H}. \tag{4.17}$$

In the plane, however, the box counting dimension cannot exceed 2. Hence, the box counting fractal dimension of an fBm trajectory in the plane is given by the expression

$$D_B = \min\left(\frac{1}{H}, 2\right) \tag{4.18}$$

compare this to expression (4.9b) for fBm trace functions. Thus, antipersistent fBms in the plane, i.e. those with Hurst exponents less than 0.5, have dimensions equal to two and will eventually fill up the 2D plane that they wander around. Persistent fBms have box counting dimensions less than two but greater than unity and are not plane filling (they would rather wander off than hang about to fill up the plane!). It is interesting to note that particle paths in turbulent fluids are often persistent with Hurst exponents greater than 0.5. This is a consequence of the large scale correlations which exist in the flow field. Recent surveys of satellite tracked drifter trajectories on the ocean surface have found that they behave as fBms in the plane with Hurst exponents around 0.78.

Trajectories in 3D space may be generated in a similar manner to the planar trajectories described above by simply using three fBm traces for the co-ordinate positions of the wandering particle. In this way, 3D fractional diffusion may be modelled. Extending this further, we may generate fBm trajectories in higher dimensional spaces. The fractal dimension of the resulting fBm flight in a D_E-dimensional Euclidean space is given by

$$D_B = \min\left(\frac{1}{H}, D_E\right). \tag{4.19}$$

In order to be volume filling an fBm in 3D space is required to be antipersistent with a Hurst exponent less than 0.333.... .

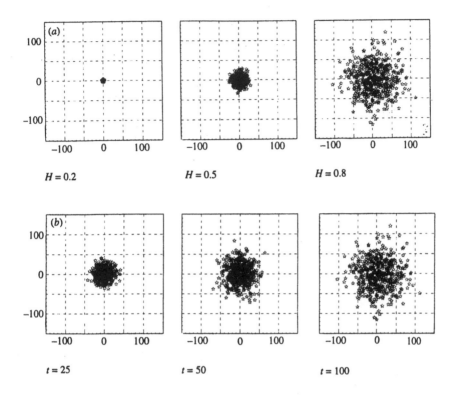

Figure 4.14. The evolution of particle clouds. (*a*) Diffusion coefficient, $D_f = 1.0$; elapsed time, $t = 100$. (*b*) Diffusion coefficient, $D_f = 1.0$; Hurst exponent, $H = 0.8$.

Diffusion processes of scientific and engineering interest, for example the diffusion of contaminants in fluids, are sometimes simulated using a particle tracking method, whereby the diffusing cloud is represented by a large number of marked particles undergoing Brownian motions. (Another popular method is to numerically solve an advection–diffusion type equation.) Trajectories of the particles in particle tracking models are generated using equation (4.1) (or equations (4.13*a, b*) for fBm) for each of the co-ordinates of each particle. Very large numbers of particles, usually of the order of 10 000, are used to simulate the dynamics of the diffusion process under observation. The diffusive behaviour of a particle cloud is shown in figure 4.14, where the spreading of 100 particles is shown. The particles are shown as hollow stars in the figure. In the plots of figure 4.14, only the final position of each particle is plotted—not the whole trajectory as before. The small number of particles used for each cloud allows us to better resolve the spreading. In figure 4.14(*a*), three particle clouds have been left to diffuse over a period of 100 s with a diffusion coefficient of 1.0. The enhancement of the spreading of the cloud with increasing Hurst exponent is evident from the plots. The Hurst exponent is kept constant at 0.8 in figure 4.14(*b*), and the particle locations at three times since release, 25, 50 and 100 s, are plotted.

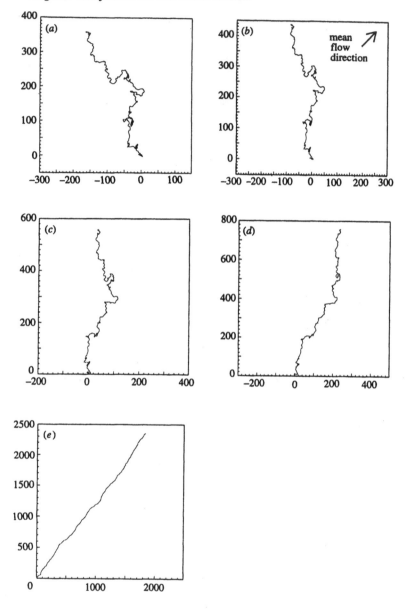

Figure 4.15. Simulated fBm particle trajectories with various mean motion components. (*a*) $U, V = 0$. (*b*) $U, V = 0.2$. (*c*) $U, V = 0.5$. (*d*) $U, V = 1.0$. (*e*) $U, V = 5.0$. 400 steps are taken in each direction with $K_f = 1.0$, $H = 0.78$ and $\Delta t = 1$. All trajectories begin from the origin. The horizontal and vertical advective velocities, U and V, are set equal to each other to produce a mean flow towards the top right of each figure. (After Addison (1996). Reproduced by permission of the IAHR.)

In practice, the diffusion of a particle cloud is superimposed onto a background flow pattern which causes an underlying mean movement of the particles called advection. The effect of increasing the underlying advection current on a single particle path is shown in figure 4.15. At each time increment the Brownian particle takes both a random diffusive step and a non-random advective step generated by the mean flow. The constant mean flow advection, towards the top right of the plots, increases from (*a*) to (*e*). The stretching of the particle path with increasing advection velocities is evident. If we tried to measure the fractal properties of these stretched particle paths, we would find a false lowering of the fractal dimension. D_B tends to unity with increasing stretching as the trajectory becomes more 'linear'. One must remove these advective effects before investigating the fractal properties of the trajectory. The same is true for any fractal object or process which has become distorted or masked in a similar manner. The particle tracking method is used in practice to solve diffusion problems of great geometric complexity with spatially varying advection velocities, for example the spreading of pollutant clouds around islands in the ocean, the movement of contaminants within air conditioning ducts, the progression of contaminants through groundwater aquifers, and so on.

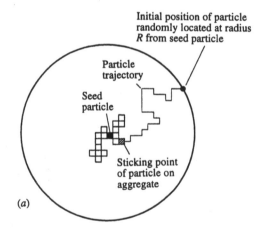
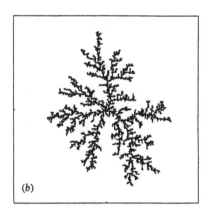

Figure 4.16. Diffusion limited aggregation. (*a*) The construction process of a DLA cluster. (*b*) A DLA cluster (generated using 5000 particles).

4.6 Diffusion limited aggregation

In this section we look briefly at a simple random growth process generated by Brownian motion known as **diffusion limited aggregation**, or DLA. This cluster growth process has been used to model a variety of physical phenomena including bacterial colonies, viscous fingering and electrochemical deposition. DLA is generated in the plane as follows.

A seed particle is located somewhere on the plane. Particles are then released from random locations far from the seed particle and allowed to follow a Brownian trajectory. Once the Brownian particle comes into contact with the seed particle it 'sticks' and a two particle cluster is formed. The process is repeated: each time the moving Brownian particle comes into contact with the cluster it sticks to it. The DLA cluster grows through the agglomeration of the diffusing Brownian particles. Figure 4.16(*a*) illustrates a practical algorithm used in the simulation of DLA. In this figure, a particle is released, at a radius *R*, far from the growing cluster. The particle is allowed to move in the plane, taking randomly selected discrete steps up, down, left or right, hence the trajectory is essentially Brownian at large scales. Once the wandering particle comes into contact with the cluster at the sticking point, it attaches itself to it. If the particle wanders away from the cluster by a significant distance (a specified multiple of *R*) it is discounted and another particle is released.

The diffusing particles may stick to any part of the cluster, however, they are more likely to encounter the tips of the cluster than to penetrate deep into its inner regions. This causes particle branches, or dendrites, to be formed, which grow outwards from the initial location of the seed particle. Figure 4.16(*b*) contains a DLA cluster formed using the above algorithm. It is a straightforward matter to extend DLA to higher spatial dimensions. The fractal dimension of DLA clusters in the plane has been estimated at D_H = 1.7. However, an analytical value for the fractal dimension of DLA clusters is currently unavailable due to the non-local growth mechanisms involved in their construction. DLA clusters constructed in 3D have a fractal dimension of 2.5.

4.7 The colour and power of noise

Many random functions or noises found in science and nature have power spectra with very definite **power law** relationships, the power spectra being a plot of the squared magnitude of the Fourier transform against frequency. This last section briefly outlines some common relationships for fBms and other random functions. Frequency and power spectra are dealt with in more detail in section 7.3, and it may be useful to reread the current section after chapter 7 has been covered.

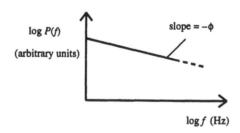

Figure 4.17. Power law scaling of coloured noise functions.

The time trace of a random variable with a Gaussian distribution (figure 4.2(*a*)) is known as white noise. The power spectral density, $P(f)$, of white noise remains constant over the frequency range, f, that is, it is proportional to $1/f^0$. The integral of white noise gives regular Brownian motion (also known as Brown noise) with a Hurst exponent of 0.5. The spectral density of Brownian motion decreases with increasing frequency and is in fact proportional to $1/f^2$. White noise and Brown noises are often observed in nature. Pink noise is defined as having a spectral density proportional to

$1/f^1$. Pink noises are found in a variety of physical systems. Finally, noises which have spectral densities proportional to $1/f^\phi$, where ϕ is greater than 2, are commonly referred to as black noises. A logarithmic plot of the power spectral density against frequency allows the power scaling exponent, ϕ, to be measured from the slope of the plot (see figure 4.17). The power law exponent of an fBm trace is related to its fractal dimension by the following relationship:

$$D_B = \left[\frac{5}{2} - \frac{\phi}{2} \right]. \tag{4.20}$$

Comparing the above expression with equation (4.9b) we arrive at the following relationship between the spectral exponent and the Hurst exponent:

$$\phi = 1 + 2H. \tag{4.21}$$

This gives spectral exponents in the range $1 < \phi < 3$ for fBms. The power spectrum, therefore, provides another method for the classification of fBms.

4.8   Chapter summary and further reading

4.8.1   Chapter keywords and key phrases

Brownian motion	*trajectories*	*traces*
Hurst exponent	*self-affinity*	*Fickian diffusion*
diffusion coefficient	*random walk*	*fractional Brownian*
persistence/antipersistence	*fractal diffusion coefficient*	* motion*
non-Fickian/anomolous diffusion	*zero set*	*Levy flight*
diffusion limited aggregation	*power laws*	

4.8.2   General

In this chapter we dealt with fractional Brownian motion (and its special case of regular Brownian motion). We saw that fractional Brownian motions, fBms, are random fractals with statistical self-similarity over all scales. The fBm is characterized by its **Hurst exponent** which is directly related to its fractal dimension. Hurst (1951, 1956) developed rescaled range analysis to determine the power law scaling of reservoir flood statistics (see also Mandelbrot and Wallis (1969d)). More information on rescaled range analysis as a technique for determining the Hurst exponent of time series data, including fBms, can be found in the book by Feder (1988). As with chapters 2 and 3, it is recommended that a good place for the reader to initiate a literature search is with the book by Mandelbrot (1982a), where another interesting fractal trace function, the Mandelbrot–Weierstrass function, is to be found.

   Although observed by Robert Brown in 1827, it was not until 1905 that a mathematical theory was produced by Albert Einstein (1926) which predicted Brownian particle movements; and it is this theory that is now known as Brownian motion. Brownian motion is composed of statistically independent steps, i.e. each step does not depend on any of the previous steps. Each step taken in an fBm depends upon the history of the fBm up to that point (Mandelbrot and Van Ness 1968). The accurate generation of an fBm from expressions (4.13a, b) requires an appropriate choice of **time increment** and **memory** (Mandelbrot and Wallis 1969a–c). In general, a large memory and small time step is required. In addition, the approximation of the integral of equation (4.12) by equations (4.13a, b) leads to a slight difference in the diffusion coefficient realized, and a slight modification is required for practical use. **Fast, accurate algorithms** have been developed for the generation of fBms (Mandelbrot 1971, Rambaldi and Pinazza 1994, Binder 1994). Many modern fBm generation algorithms use **spectral methods** (for good examples see the work of Turcotte (1992) or Yin (1996)), however, these methods are really outside the scope of this elementary text where we have used the more obvious discrete integration described by equations (4.13a, b). Mandelbrot (1984b) summarizes the many fractal dimensions created by fBm traces and trajectories in various spaces. In this chapter we used the box counting dimension, D_B, to find a fractal dimension estimate for the traces and trajectories of fBm. It is worth noting that the divider dimension D_D may also be used to find a fractal dimension estimate similar to D_B for fBm trajectories (see Q4.6 and Q4.7 in the next section). However, using the structured walk technique to find a fractal dimension for self-affine fBm time traces will lead to a dimension estimate, D_D, between one and $1/H$, which depends upon the relative scale of the temporal and spatial axes. You may like to try and prove this for yourself noting that the step size λ is a function of both time and space, otherwise see the book by Feder (1988). Talibuddin and Runt (1994) investigate the reliability of a range of popular techniques for determining the **dimension of fBm surfaces**; see also the article by Kaplan and Kuo (1995) who employ a **wavelet** based method for the same problem.

4.8.3   Diffusion

Much interest has centred on diffusion within media with their own fractal structure, where non-Fickian diffusion can result solely from the fractal structure itself (Orbach 1986, Reis 1995a, Du *et al* 1996, Wang and Cohen 1996). Havlin and Ben-Avraham (1987) give a comprehensive account of the phenomena. Diffusion through soils, which themselves have a fractal structure, is discussed by Crawford *et al* (1993) and Sahimi (1993). A fractal dimension of two has been found by Rapaport (1985) for the **simulation of a hard-sphere fluid**, corresponding to the regular Brownian motion of the particles. Kobayashi and Shimojo (1991) have found fBm trajectories for ions in **molecular dynamics** simulations; and Giaever and Keese (1989) have found the motion of **mammalian cells** to be fBms. Persistent fractional Brownian trajectories have been found by Sanderson and Booth (1991) and by Osborne *et al* (1989) for **drogued ocean surface drifters** undergoing turbulent diffusion. (Note that spreading due to turbulent diffusion may scale with exponents beyond unity (Richardson 1926, Okubo 1971, Hentschel and Procaccia 1984).) The diffusion of Brownian particles and the use of particle tracking models in environmental diffusion is outlined by Hunter *et al* (1993) and more generally in the text by Csanady (1973). Delay *et al* (1996) present a 2D Fickian, particle tracking model, and Addison (1996) and Addison and Bo (1996) give details of a simple non-Fickian fBm particle tracking model. **Fractional diffusion equations** derived from the limiting dynamics of random walks are discussed by Roman and Alemany (1994) and Compte (1996); see also the article by Wang (1994) who details the diffusion characteristics of Brownian particle movements with **added noise and friction effects**. Levy flights are discussed in a variety of contexts by Shlesinger (1989). The use of fractal diffusion and Levy statistics in **biology** is reviewed by West (1995); and specifically in **physiology** by West and Deering (1994). Miscellaneous papers of interest dealing with **fBms** and **anomalous diffusion** include those by: Eisenberg *et al* (1994), Burioni and Cassi (1994), Zumofen and Klafter (1994), Ding and Yang (1995), Kotulski (1995), Reis (1995b) and Gripenberg and Norros (1996).

4.8.4 fBm landscapes etc

Extensive use has been made of fBm trace functions to model **surfaces and landscapes**: see for example the collection of articles by Peitgen and Saupe (1988). Their text gives additional methods (and computer algorithms) for the generation of fBms including the **displacement methods** and **spectral synthesis**. The fractional Brownian landscapes of figure 4.11 were generated using an algorithm from this text; see also the article by Voss (1985). More details on **alternative methods of fBm generation** can be found in Kreuger *et al* (1996), Kaplan and Kuo (1996), and Yin (1996). Mandelbrot (1982b) and Fournier *et al* (1982) discuss the **computer rendering** of fractal landscapes. Pumar (1996) discusses the use of **fractal interpolation** in **terrain imagery**. General information on **fractals and computer graphics** is to be found in the text by Encarnaçao *et al* (1992). An interesting article by Herzfeld *et al* (1995) asks the question, '**Is the Ocean Floor a Fractal?**', and a comprehensive review of **fractals and landscapes** is given by Xu *et al* (1993). The application of fractal image analysis to the earth's topography is discussed by Huang and Turcotte (1990a). Self-affinity in **music, mountains** and **clouds** is considered by Voss (1989). Liu and Molz (1996) find fBm properties for **ground permeabilities**. An interesting article by Collins and De Luca (1994) documents the **random walk** which the human body undergoes during standing. Peng *et al* (1992), Buldyrev *et al* (1993), Voss (1993), and Arneodo *et al* (1996) discuss the fBm properties of **DNA sequences**. Antipersistent fBm has been described by Cioczek-Georges and Mandelbrot (1995) for a class of micropulses.

4.8.5 DLA

Diffusion limited aggregation (Witten and Sander 1981) has been used extensively to describe various aggregation processes in nature including: **bacterial colonies** (Ben-Jacob *et al* 1994); **dielectric breakdown** (Niemeyer *et al* (1984); **electrochemical deposition** (Grier *et al* 1986); **viscous fingering** (Nittman *et al* 1985, Feder *et al* 1989), **material failure** (Meakin 1991); **chemical dissolution** (Daccord 1987); **neuron development** (Caserta *et al* 1990); **nanometre scale surface structures** (Röder *et al* 1993); **growing interfaces within porous media** (Ferer and Smith 1994); and **aluminium–silica chemical reactions** (Balazs *et al* 1996). Meakin (1983) gives estimates of the fractal dimension of DLA clusters grown in 2 to 6D space. The geometrical complexity of DLA is probed by Argoul *et al* (1989) using **wavelet transforms**. DLA is put forward, together with other aggregate growth models, as a possible mechanism for the generation of **flame generated soot materials** by Megaridis and Dobbins (1990). The texts by Vicsek (1989) and Barabasi and Stanley (1995) contain more detail on **fractal growth phenomena**, see also the collection of articles by Stanley and Ostrowsky (1986). The text by Charmet *et al* (1990) devotes about one quarter of its contents to the use of DLA in describing **disorder** and **fracture**. Stanley (1989) gives some thoughts on experiments involving DLA. The reader is also referred to many of the texts cited at the end of chapter 3, where more information on fractal diffusion processes can be found, i.e. **chromatograph fronts**, **diffusion fronts** and **viscous fingering**.

4.8.6 Power spectra

For a discussion of **power spectra** and associated fractal dimensions of time series the reader is referred, in the first instance, to the articles by Osborne and Provenzale (1989) and Higuchi (1990). A more general form of equation (4.19), which covers the power spectral exponent for N-variable fBm trace functions, can be found in the book by Peitgen and Saupe (1988). Reed *et al* (1995) provide a more mathematical treatment of the power law relationships, and Nakagawa (1993) details a method to obtain the fractal dimension based on the moment of the power spectrum. Power law relationships have been used by Hasegawa *et al* (1993) in the **characterization of engineering surfaces**, and by Jun-Zheng and Zhi-Xiong (1994) for the determination of the fractal dimension of **soil strengths**. Issa and Hammad (1994) use 2D power spectra to find fractal dimensions for **fractured concrete surfaces**. In addition, they use the perimeter–area relationship (section 3.6) to find the fractal dimension. They do this by using the slit-island technique which essentially sections the surface as various levels to produce contours (i.e. zerosets on a plane). D_B is then found for the closed contours (fractal islands) using the perimeter–area relationship. Hence D_B for a surface can be found, knowing that it exceeds the zeroset dimension by one. Mandelbrot *et al* (1984) use both the slit-island technique and frequency spectra to estimate the fractal dimension of fractured steel specimens, which they then relate to impact energy.

4.9   Revision questions and further tasks

Q4.1 List the keywords and key phrases of this chapter and briefly jot down your understanding of them.

Q4.2 What is the dimension of an fBm trace function with a Hurst exponent of 0.4? What is the dimension of its zeroset?

Q4.3 Using the random numbers provided in table Q4.3 below generate the first 50 steps of a Brownian motion time trace. (Use the first two columns.)

Q4.4 Again, using the random numbers provided in table Q4.3 below generate the first 50 steps of a Brownian motion trajectory in a plane. (This time use all four columns.)

Q4.5 If the fractal dimension of a surface, simulated using a two-variable fBm trace function (i.e. an fBm surface such as those in figure 4.11), is known to be 2.32, what is the corresponding Hurst exponent?

Q4.6 Figure Q4.6 shows a blow up of the fBm trajectory plotted in figure 4.12(*c*). Find its fractal dimension using the box counting method.

Q4.7 Find the divider dimension of the fBm trajectory in figure Q4.6 using the structured walk technique.

Table Q4.3. 100 random numbers taken from a Gaussian distribution with zero mean and unit standard deviation.

0.430 68	−1.376 922	0.376 201	0.545 058
−0.026 185	1.640 878	−0.789 929	0.762 248
−1.649 632	−0.136 708	−2.322 634	0.727 9
0.075 367	1.306 08	−0.717 792	−0.163 546
1.049 633	1.123 124	−0.401 758	−0.622 066
0.644 617	0.767 721	1.184 539	2.145 141
0.372 513	−0.060 2	0.384 111	2.125 531
0.151 137	−0.072 975	−0.648 226	0.726 776
−0.259 026	−1.294 886	−0.217 875	−0.799 359
0.951 021	0.988 462	1.868 09	0.711 032
−0.723 172	−0.194 835	0.444 018	0.293 716
−0.987 809	0.325 155	−0.752 387	−0.359 428
1.108 488	−0.289 098	−0.298 055	0.906 802
0.519 599	−0.045 219	0.487 022	1.268 712
−0.488 468	1.785 714	0.635 386	0.209 443
−0.222 429	−0.947 07	0.060 896	−0.048 845
−1.166 355	−1.069 849	−1.342 955	1.739 85
0.697 033	1.114 764	−0.501 965	0.381 856
−2.302 259	−1.192 579	−0.514 155	0.644 689
0.488 623	−0.407 272	−0.254 251	−2.320 918
−0.831 877	1.305 98	0.107 032	0.141 534
−1.066 651	2.067 958	−0.173 012	0.288 215
−1.474 877	−0.340 118	−1.323 482	−0.194 752
−0.256 335	−1.116 14	−0.717 592	0.151 686
1.041 411	−0.037 99	0.742 781	−0.591 089

Q4.8 Write a computer program to generate Brownian time traces and Brownian trajectories in 2D and 3D space. (Essentially computerize the method of Q4.4 using the random number generator of your computer.) Verify visually the self-affinity property of Brownian time traces by rescaling portions of your generated trace.

Q4.9 (*a*) Using the time series output from the program you developed in Q4.8, verify the scaling relationship given by expression (4.2) in the text.

(*b*) Using a large number of time series Brownian motion traces, verify the scaling relationships for a cloud of Brownian particles given by expressions (4.3*a–d*).

Q4.10 Write a computer program to generate fBm traces and trajectories in 2D and 3D space, using the method outlined in section 4.3, equations (4.13*a*, *b*), and figure 4.12.

Q4.11 (*a*) Using the time series output from the program you developed in Q4.10, verify the scaling relationship given by expression (4.4) in the text for $H = 0.8$.

(*b*) Using a large number of time series fBm traces, verify the scaling relationship given by expression (4.5*a*) for a cloud of fractional Brownian particles, again use $H = 0.8$.

Q4.12 (*a*) Modify your Brownian motion program of Q4.8 to produce the zeroset of the time trace.

(*b*) Use the zeroset of (*a*) to generate a Levy flight.

Q4.13 Following the method outlined in section 4.6 and figure 4.16, write a program to produce DLA clusters in the plane.

Q4.14 The power spectral exponent of a known fBm function is found to be 2.6. What is the corresponding fractal dimension and Hurst exponent of the function?

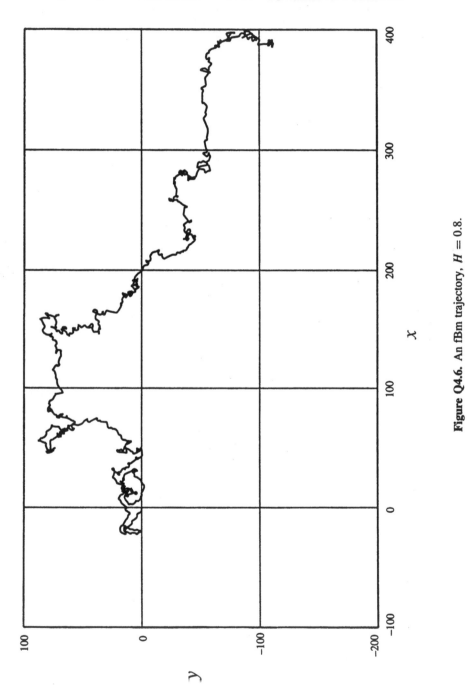

Figure Q4.6. An fBm trajectory, $H = 0.8$.

Chapter 5

Iterative feedback processes and chaos

5.1 Introduction

In this chapter we will investigate **discrete dynamical systems**. These systems evolve through a series of discrete steps in time, and not continuously, in contrast to the continuous dynamical systems which we will deal with in chapters 6 and 7. Discrete dynamical systems evolve through time by the process of iteration, where the subsequent state of the system is determined by its present state. As a simple example, consider a 1D dynamical system whose state is given by the variable x where an updated value of x is produced solely from its present value. To put it another way, the value of x at a subsequent time $(n + 1)$ is a function only of its present state at time n. This is written as

$$x_{n+1} = f(x_n). \tag{5.1}$$

This type of expression is known as an iterated map as it maps the present value of the variable x at time n onto the next value at time $n + 1$. The repeated iteration of relatively simple equations of this form may lead to both simple and complex model behaviours. In this chapter we will observe some of the rich variety of behaviours that iterative processes may produce.

5.2 Population growth and the Verhulst model

The simplest model of population growth assumes a constant growth rate whereby the population increases in size by a fixed proportion in each time interval. This is written as,

$$P_{n+1} = P_n + r'P_n = (1 + r')P_n = rP_n \tag{5.2}$$

where P_{n+1} is the population size at time $n + 1$, P_n is the population size at time n, r' is the growth rate (the same as interest rate in compound interest calculations), and r is the growth factor $= 1 + r'$.

It follows from equation (5.2) that $r = (P_{n+1} - P_n)/P_n = \Delta P/P_n$—usually expressed as a percentage.

Then

$$P_n = (1 + r')^n P_0 = r^n P_0 \tag{5.3}$$

where P_0 is the initial population at time $t = 0$. However, in this simple model (with r constant) the population grows without bounds, i.e. as n tends to infinity so does the population, P. Clearly this does not happen for real populations, which are limited in their growth by external factors such as food, space, disease, etc. Therefore, there must exist a maximum possible population size $P_{\max}$. To take this maximum population size into account in a population model, a modified growth rate, r', is used, which is proportional to the difference between the population size at time n (i.e. P_n) and the maximum possible population size $P_{\max}$, expressed as follows:

$$r' = a(P_{\max} - P_n) \tag{5.4}$$

where a is a constant. Now, when $P_n = P_{\max}$, it follows that $r' = 0$ and no further growth takes place. Substituting equation (5.4) into equation (5.2) leads to

$$P_{n+1} = P_n + a(P_{\max} - P_n)P_n \tag{5.5a}$$

and this is known as the **Verhulst model** of population growth.

If we expand equation (5.5a) we obtain

$$P_{n+1} = P_n + aP_{\max}P_n - aP_n^2 \tag{5.5b}$$

and we see that the last term provides a negative **nonlinear feedback** to the equation. It is this term which causes the Verhulst model to posses a rich variety of behaviour including chaotic motion. Often, it is more usual to use relative population sizes (denoted by lower case p) where the populations are expressed as a fraction of the maximum possible population size. Equation (5.5a) then becomes

$$p_{n+1} = p_n + ap_n(1 - p_n) \tag{5.6}$$

where

$$p_{n+1} = \frac{P_{n+1}}{P_{\max}} \qquad p_n = \frac{P_n}{P_{\max}} \qquad p_{\max} = \frac{P_{\max}}{P_{\max}} = 1.$$

5.3 The logistic map

Equation (5.6) is the Verhulst model normalized by the maximum population size. In this, and the next four sections, we will concentrate on a slightly simpler cousin of the Verhulst equation known as the **logistic map**. To obtain this map, we simply remove the first term from equation (5.6). In its more usual form, in terms of x instead of p, the logistic map is written as

$$x_{n+1} = ax_n(1 - x_n) \tag{5.7a}$$

where the current value of the variable x, i.e. x_n, is mapped onto the next value, i.e. x_{n+1}. The logistic map is a **non-invertible** map. Non-invertible means that although we may iterate the map forward in time with each x_n leading to a unique subsequent value, x_{n+1}, the reverse is not true, i.e. iterating backwards in time leads to two solutions for x_n for each value of x_{n+1}. By repeatedly iterating the logistic map forward through time, we may observe various behaviours of the iterated solutions. The sequence of iterated solutions to the map is called an **orbit**. The behaviour of successive iterates of the logistic map depends both on the **control parameter**, a, and the initial condition (or starting point) used for the iterations. The function corresponding to the logistic map (equation (5.7b)), known as the logistic function or the **logistic curve**, is the parabolic curve containing all the possible solutions for x in equation (5.7a).

$$f(x_n) = ax_n(1 - x_n). \tag{5.7b}$$

The logistic curve is a member of a large family of iterated map functions known as unimodal (single humped) map functions which are smooth curves with single maxima (here the single maximum is at $x = \frac{1}{2}$). In section 5.6 we shall use the logistic curve to graphically iterate the map of equation (5.7a).

5.4 The effect of variation in the control parameter

In this section, we will produce successive iterated solutions, or orbits, to the logistic map (equation (5.7a)) for five values of the control parameter a. At each step in the iteration, the current value of x is fed back into the map to produce the next value, then this value is used to produce the next and so on, that is:

$$\begin{aligned} x_1 &= ax_0(1 - x_0) \\ x_2 &= ax_1(1 - x_1) \\ x_3 &= ax_2(1 - x_2) \quad \text{etc} \end{aligned} \tag{5.7c}$$

where the initial value x_0 is known. This feedback process is shown schematically in figure 5.1. In each of the following cases, the initial value of x, i.e. x_0, is set to 0.2.

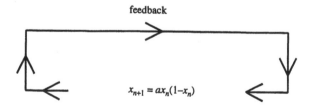

Figure 5.1. The iterated feedback process of the logistic map.

5.4.1 $a = 0.90$

Successive iterations of the logistic map (5.7a) for the control parameter $a = 0.90$ produce the following values of x:

$x_0 = 0.2$
$x_1 = 0.1440$ $x_2 = 0.1109\ldots$ $x_3 = 0.0887\ldots$ $x_4 = 0.0727\ldots$ $x_5 = 0.0607\ldots$
$x_6 = 0.0513\ldots$ $x_7 = 0.0438\ldots$ $x_8 = 0.0377\ldots$ $x_9 = 0.0326\ldots$ $x_{10} = 0.0284\ldots$
$x_{11} = 0.0248\ldots$ $x_{12} = 0.0218\ldots$ $x_{13} = 0.0192\ldots$ $x_{14} = 0.0169\ldots$ $x_{15} = 0.0150\ldots$
$x_{16} = 0.0133\ldots$ $x_{17} = 0.0118\ldots$ $x_{18} = 0.0105\ldots$ $x_{19} = 0.0093\ldots$ $x_{20} = 0.0083\ldots$
etc.

(Note that, herein, three dots after the number is taken to mean that further non-zero decimal places exist, but not necessarily extending to infinity.)

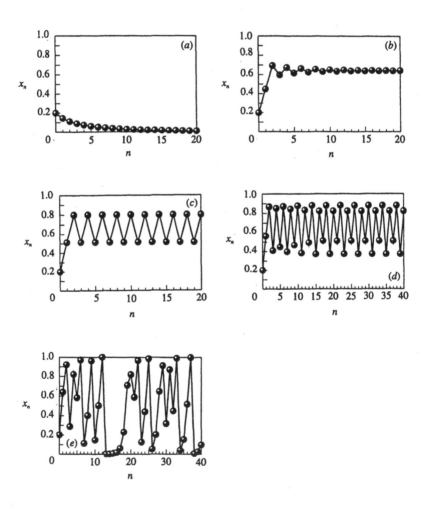

Figure 5.2. Iterated solutions of the logistic equation plotted for selected values of the control parameter a. (*a*) $a = 0.9$ (decay to zero). (*b*) $a = 2.60$ (steady state period 1). (*c*) $a = 3.20$ (period 2). (*d*) $a = 3.52$ (period 4). (*e*) $a = 4.00$ (chaotic).

The iterations seem to be heading towards zero. In figure 5.2(a), the first twenty iterated solutions to the logistic map, x_n, are plotted against n. The figure is the time history, or **time series** of the iterated solutions to the logistic map, with x plotted at each discrete time step n. Straight line segments are used to connect the iterations for the purpose of visualization only. Remember that the system is not continuous in time, but rather discrete, and hence the line segments do not represent a path through time from one value of x to the next. In fact, for the control parameter $a = 0.90$ and an initial condition of $x_0 = 0.2$, the orbit of the logistic map tends asymptotically towards zero as n tends to infinity. Indeed, all values of a between zero and one lead to iterations which decay towards zero as n increases.

5.4.2 $a = 2.60$

Successive iterations of the logistic map for the control parameter $a = 2.60$ produces the following values of x:

$x_0 = 0.2$				
$x_1 = 0.4160$	$x_2 = 0.6316\ldots$	$x_3 = 0.6049\ldots$	$x_4 = 0.6213\ldots$	$x_5 = 0.6117\ldots$
$x_6 = 0.6175\ldots$	$x_7 = 0.6140\ldots$	$x_8 = 0.6161\ldots$	$x_9 = 0.6149\ldots$	$x_{10} = 0.6156\ldots$
$x_{11} = 0.6152\ldots$	$x_{12} = 0.6154\ldots$	$x_{13} = 0.6153\ldots$	$x_{14} = 0.6154\ldots$	$x_{15} = 0.6153\ldots$
$x_{16} = 0.6153\ldots$	$x_{17} = 0.6153\ldots$	$x_{18} = 0.6153\ldots$	$x_{19} = 0.6153\ldots$	$x_{20} = 0.6153\ldots$
etc.				

The iterations seem to have settled down to a repeating solution of $0.6153\ldots$. Notice that the above iterated values are correct to within one decimal place of the final solution after two iterations (i.e. $0.6\ldots$), correct to two decimal places after five iterations (i.e. $0.61\ldots$), and correct to three decimal places after 10 iterations (i.e. $0.615\ldots$). After fifteen iterations the solution has settled down to $0.6153\ldots$. In general, higher resolution (accuracy) of the final solution requires more iterations. In figure 5.2(b), the first twenty iterated solutions to the logistic map, x_n, are plotted against n for $a = 2.60$ and $x_0 = 0.2$. We see from the figure that the logistic map settles down, through a decaying oscillation, to the single value of $0.6153\ldots$. This final solution is known as a **period 1 orbit**, as the iterates tend to a fixed value where $x_{n+1} = x_n$ for large n.

5.4.3 $a = 3.20$

Successive iterations of the logistic map for the control parameter $a = 3.20$ produces the following sequence of solutions:

$x_0 = 0.2$				
$x_1 = 0.5120\ldots$	$x_2 = 0.7995\ldots$	$x_3 = 0.5130\ldots$	$x_4 = 0.7994\ldots$	$x_5 = 0.5130\ldots$
$x_6 = 0.7994\ldots$	$x_7 = 0.5130\ldots$	$x_8 = 0.7994\ldots$	$x_9 = 0.5130\ldots$	$x_{10} = 0.7994\ldots$
$x_{11} = 0.5130\ldots$	$x_{12} = 0.7994\ldots$	$x_{13} = 0.5130\ldots$	$x_{14} = 0.7994\ldots$	$x_{15} = 0.5130\ldots$
$x_{16} = 0.7994\ldots$	$x_{17} = 0.5130\ldots$	$x_{18} = 0.7994\ldots$	$x_{19} = 0.5130\ldots$	$x_{20} = 0.7994\ldots$
etc.				

In this case, the solutions rapidly converge to two alternating attracting fixed points of 0.5130... and 0.7994.... These solutions to the logistic map repeat every second value, i.e. $x_{n+2} = x_n$. This is known as a **period 2 orbit**. The period 2 orbit is shown graphically in figure 5.2(c) where the first twenty iterated solutions to the logistic map are plotted against n.

5.4.4 $a = 3.52$

Successive iterations of the logistic map for the control parameter $a = 3.52$ produces the following sequence of solutions:

$x_0 = 0.2$

$x_1 = 0.5632$	$x_2 = 0.8659...$	$x_3 = 0.4086...$	$x_4 = 0.8506...$	$x_5 = 0.4472...$
$x_6 = 0.8702...$	$x_7 = 0.3975...$	$x_8 = 0.8430...$	$x_9 = 0.4657...$	$x_{10} = 0.8758...$
$x_{11} = 0.3826...$	$x_{12} = 0.8315...$	$x_{13} = 0.4930...$	$x_{14} = 0.8798...$	$x_{15} = 0.3721...$
$x_{16} = 0.8224...$	$x_{17} = 0.5139...$	$x_{18} = 0.8793...$	$x_{19} = 0.3735...$	$x_{20} = 0.8237...$

etc.

We see that the above iterates have not noticeably converged by the twentieth iteration. However, iterating further produces a **period 4 orbit**, $x_{n+4} = x_n$ (to a resolution of four decimal places by the fortieth iteration), e.g.

$x_{37} = 0.5120...$	$x_{38} = 0.8794...$	$x_{39} = 0.3730...$	$x_{40} = 0.8233...$
$x_{41} = 0.5120...$	$x_{42} = 0.8794...$	$x_{43} = 0.3730...$	$x_{44} = 0.8233...$

etc.

Figure 5.2(d) plots the first forty iterations of the logistic map with control parameter $a = 3.52$. The period 4 nature of the time series is evident from this figure.

5.4.5 $a = 4.00$

Finally, iterating the logistic map for the control parameter $a = 4.00$ produces the following solutions:

$x_0 = 0.2$

$x_1 = 0.6400$	$x_2 = 0.9216$	$x_3 = 0.2890...$	$x_4 = 0.8219...$	$x_5 = 0.5854...$
$x_6 = 0.9708...$	$x_7 = 0.1133...$	$x_8 = 0.4019...$	$x_9 = 0.9615...$	$x_{10} = 0.1478...$
$x_{11} = 0.5038...$	$x_{12} = 0.9999...$	$x_{13} = 0.0002...$	$x_{14} = 0.0009...$	$x_{15} = 0.0038...$
$x_{16} = 0.0153...$	$x_{17} = 0.0605...$	$x_{18} = 0.2275...$	$x_{19} = 0.7031...$	$x_{20} = 0.8348...$

etc.

The above iterations have not noticeably converged onto a periodic orbit by the 20th iteration and, unlike the $a = 3.52$ case above, further iterations will never produce a repeating, and hence periodic, sequence of solutions. This aperiodic behaviour is known as a **chaotic orbit** or simply **chaos**. The first forty iterations are shown in figure 5.2(e). Notice that an iterated value, x_n, can only occur once, since if one value repeated itself then, due to the deterministic nature of the iterations, the next value must be a repetition, and the next and so on.

5.5 The general form of the iterated solutions of the logistic map

When iterating the logistic map, the iterated solutions eventually settle down to a final behaviour type. The sequence of iterated solutions produced by the initial iterations is known as the transient orbit of the system. The final sequence of iterated values that the iterations tend to is known as the post-transient orbit. Transient behaviour is obvious in figure 5.2(*a*) where the orbit tends asymptotically to zero. The transient orbit is also noticeable in figure 5.2(*b*) as a decaying oscillation, especially over the first ten or so iterations, however, by the twentieth iteration there is no noticeable change in the iterated solutions. In fact the orbit tends asymptotically to the fixed point of 0.6153.... In figure 5.2(*c*), the transient behaviour seems to disappear rather fast with only the initial condition appearing to be noticeably 'off' the post-transient period 2 orbit. In figure 5.2(*d*), a post-transient period 4 orbit seems to be well established by the tenth iteration. Transient behaviour is not at all noticeable in figure 5.2(*e*) due to the erratic nature of the post-transient, chaotic orbit. This is quite often the case for chaotic motion where the complex structure of the post-transient behaviour may make it difficult to observe the initial transient behaviour. In general, the post-transient orbits are approached asymptotically by the iterates of the logistic map. In practice, a finite resolution (i.e. accuracy to within a given number of decimal places) of the post-transient orbit is usually all that is required. The resolution obtained is directly related to the number of iterations undertaken. This was illustrated in the above section for the $a = 2.6$ case. Note that if one starts the iteration procedure on a post-transient orbit then no transient orbit will occur.

    The orbit of the logistic map is then attracted, through repeated iteration, towards the post-transient orbit, whether it is periodic or chaotic. The **attractor** is the set of points approached by the orbit as the number of iterations increases to infinity. The post-transient sequences of iterated solutions to the map lie on the attractor. If the system settles down to a periodic orbit, then the system is said to have a **periodic attractor**, e.g. a period 1 attractor, period 2 attractor or period 4 attractor as seen above. If, on the other hand, the system behaves chaotically with an aperiodic orbit, then the system is said to have a **chaotic attractor**, more commonly referred to as a **strange attractor**. The logistic map with $a = 4.0$ has a strange attractor.

5.6 Graphical iteration of the logistic map

Successive iterates of the logistic map may be found graphically using a plot of the logistic curve (equation (5.7*b*)) and a 45° line corresponding to $x_n = x_{n+1}$. This process is illustrated in figure 5.3(*a*). We begin on the horizontal axis with an initial value of x_0 and project upwards from this point to the logistic curve. The point at which this projection hits the curve enables the next value of x, i.e. x_1, to be read by projecting horizontally onto the vertical, x_{n+1} axis (shown by the dashed line in the figure). In order to find the second iterate, we make $x_1 = x_n$ and place it onto the horizontal axis.

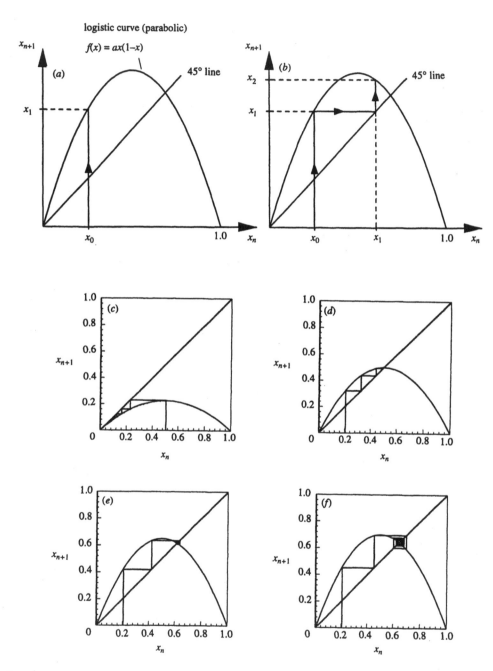

Figure 5.3. Graphical iteration of the logistic equation. (*a*) Finding x_1 from x_0. (*b*) Finding x_2 from x_1. (*c*) $a = 0.90$; $x_0 = 0.5$ (period 1). Fixed point at origin. (*d*) $a = 2.00$ (period 1). Fixed point at logistic curve maximum $= 0.5$. (*e*) $a = 2.60$ (period 1). (*f*) $a = 2.80$ (period 1). (*g*) $a = 3.20$ (period 2). (*h*) $a = 3.20$; $x_0 = 0.7$ (period 2). (*i*) $a = 3.42$ (period 2). (*j*) $a = 3.52$ (period 4). (*k*) $a = 3.52$ (period 4—post transient). (*l*) $a = 4.00$ (chaos). (Continued overleaf.)

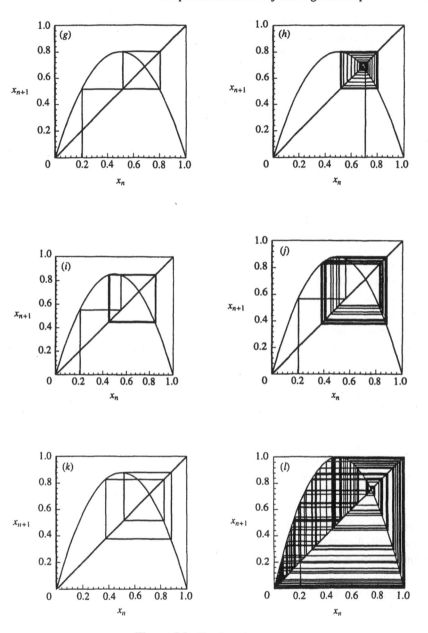

Figure 5.3. Continued.

To do this, the value of x_1, currently read from the vertical axis, is projected horizontally onto the 45° line, then downwards on to the horizontal axis. Now we have x_1 on the x_n axis. To find x_2 from x_1 we simply repeat the procedure employed to find x_1 from x_0. This is shown in figure 5.3(*b*). This graphical procedure is then repeated over and over to iterate the map.

Figures 5.3(*c–l*) contain various graphical iterations of the logistic map, some of which are directly related to the orbits of figure 5.2. Figure 5.3(*c*) contains the graphical iterations for $a = 0.9$. The attraction of the iterates towards the origin is obvious in the figure: compare this with figure 5.2(*a*). (Notice that, for reasons of clarity, the iterations in figure 5.3(*c*) were started on $x_0 = 0.5$ and not 0.2 as in figure 5.2(*a*).) All iterations of the logistic map for a between zero and one lead to orbits which head towards the fixed point at the origin, i.e. $x_n = x_{n+1} = 0$. This is true for all initial conditions, x_0, between zero and one. The graphical iteration of the logistic map for $a = 2.0$ is shown in figure 5.3(*d*). At this value of the control parameter, the 45° line intersects the parabola at its maximum. All values of a between one and two produce period 1 attractors to the left of the maximum. Values of a between two and three produce period 1 attractors to the right of the maximum. In figure 5.3(*e*) the graphical iteration is shown for $a = 2.60$ (compare with figure 5.2(*b*)). The iteration is started on $x_0 = 0.2$, and the orbit quickly converges to the steady state fixed point attractor of 0.6153.... The attractor occurs at the non-zero crossing point of the logistic map, $x_{n+1} = ax_n(1 - x_n)$, and the line, $x_{n+1} = x_n$. Substituting to eliminate the x_{n+1} term we obtain

$$x_n((a - 1) - ax_n) = 0. \tag{5.8}$$

The non-zero solution to this equation is $(a - 1)/a$. Here, $a = 2.60$ gives the fixed point as $\frac{8}{13}$ or 0.6153..., as we found by iteration above in section 5.4. The convergence of the $a = 2.80$ case in figure 5.3(*f*) is slower and the winding down of the iterated solutions to the period 1 fixed point of 9/14 or 0.6428... is more obvious. Figure 5.3(*g*) shows the period 2 orbit for $a = 3.20$. In this figure, the choice of initial condition of $x_0 = 0.2$ has lead to a rapid convergence of the solutions to the steady state period 2 orbit of 0.5130... and 0.7994.... Contrast this with figure 5.3(*h*) where the iterations are begun at $x_0 = 0.7$ and the solutions wind out from the unstable fixed point of 0.6875, where the parabolic logistic curve and the 45° line intersect. In fact, all iterations begun in the interval $(0 < x_0 < 0.6875; 0.6875 < x_0 < 1)$ end up approaching the period 2 attractor. This interval is known as the **basin of attraction** for the period 2 attractor. Iterations begun on $x_0 = 0$ or 1.0 result in fixed points of 0 and only iterations begun exactly upon the unstable fixed point of 0.6875 remain at that value. Figure 5.3(*i*) shows another period 2 orbit for the case of $a = 3.42$.

As we saw in figure 5.2(*d*), a period 4 attractor exists for $a = 3.52$. Figure 5.3(*j*) shows the approach of the iterations to the period 4 attractor from an initial condition of $x_0 = 0.2$. The period 4 attractor is obscured by the many lines of approaching iterated solutions. This is rectified in figure 5.3(*k*), where the first twenty transient iterations are not shown and the post-transient period 4 attractor is now obvious. Finally, the graphical iteration for the control parameter, a, set to 4.0 is shown in figure 5.3(*l*). In this figure only the first hundred iterations are shown for clarity as the iterations quickly fill up the plot. This map has a strange attractor and the iterations never repeat.

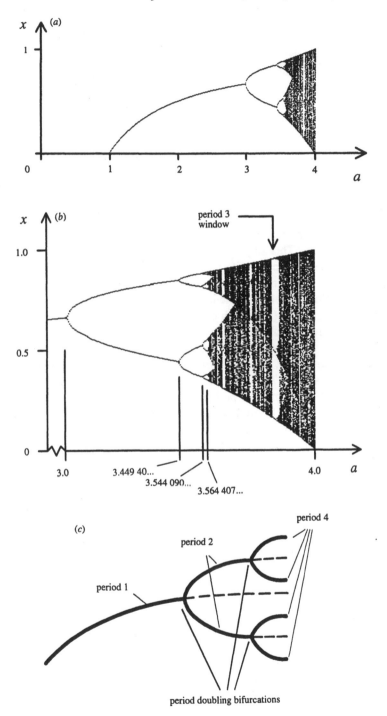

Figure 5.4. The bifurcation diagram for the logistic equation. (*a*) The bifurcation diagram for the logistic diagram: post-transient solution against control parameter. (*b*) Zoom into the logistic diagram over the range $3.0 < a < 4.0$. (*c*) General form of the period doubling bifurcations.

5.7　Bifurcation, stability and the Feigenbaum number

So far we have observed periodic attractors, of periods 1, 2 and 4, and a chaotic attractor for the logistic map. These were obtained simply by changing the control parameter a. To examine the global effect of the control parameter a on the behaviour of the logistic map it is useful to plot the post-transient solutions of the logistic map against a. Such a plot is given in figure 5.4(a). As we already know, the orbits decay to zero for values of a between 0.0 and 1.0. Values of a between 1.0 and 3.0 produce post-transient solutions to the logistic map which are period 1 fixed points. These period 1 fixed points increase in value as a is increased and form the continuous line in figure 5.4(a), rising from the a axis at $a = 1.0$ to the first **period doubling bifurcation** at $a = 3.0$. For values of the control parameter between 3.0 and 3.449 490... a period 2 attractor exists (see figure 5.4(b)). Between 3.449 490... and 3.544 090... a period 4 attractor exists, followed in sequence by attractors of period 8 (between 3.544 090... and 3.564 407...), period 16, period 32 and so on. This sequence carries on, doubling in period each time, until an infinite period is reached at a finite value of the control parameter of $a = 3.569\,945\ldots$, at which point the behaviour is chaotic and a strange attractor exists. This sequence is known as the **period doubling route** to chaos. The first few period doubling bifurcations may be seen in figure 5.4(b), which contains a blow-up of the diagram of figure 5.4(a) in the region of the control parameter between 3.0 and 4.0.

　　Between $a = 3.569\,945\ldots$ and 4.0 a rich variety of attractors exist, both periodic and chaotic. Furthermore, periodic attractors occur in this region which are not integer powers of 2. Period 3, 5, 7, 9, 11 attractors all exist. In fact, periodic attractors exist for all odd integers. For example, a period 3 attractor occurs at $a = 3.828\,435\ldots$, causing a window to open in the bifurcation diagram. This is indicated in figure 5.4(b). This period 3 attractor itself undergoes period doubling bifurcations to chaos. In fact, if we observe a period 3 in a dynamical system then this implies that we should encounter chaos in the system.

　　The period doubling sequence outlined above occurs through the bifurcation (splitting into two parts) of the previous fixed points when they become unstable. This splitting is sometimes known as a pitchfork bifurcation due to its shape (although this term is not used here in its standard mathematical sense). This is shown schematically in figure 5.4(c). At each period doubling bifurcation point, the previously stable attracting periodic fixed point becomes unstable, that is, it begins to repel iterated solutions in its vicinity, and two new stable fixed points emerge. Unstable fixed points are shown dashed in figure 5.4(c). These period doubling bifurcations give their name to the **bifurcation diagram** of figures 5.4(a) and (b).

　　Fixed points which attract the solutions to a map are known as stable; those which repel nearby solutions are unstable; and those which neither attract nor repel nearby solutions are said to be neutrally stable, or indifferent. Local stability depends on the absolute value of the slope of the map function at the fixed point, i.e.

$$\text{stable:}\quad \left| f'(x_n) \right| < 1 \tag{5.9a}$$

$$\text{neutrally stable:}\quad \left| f'(x_n) \right| = 1 \tag{5.9b}$$

$$\text{unstable:}\quad \left| f'(x_n) \right| > 1. \tag{5.9c}$$

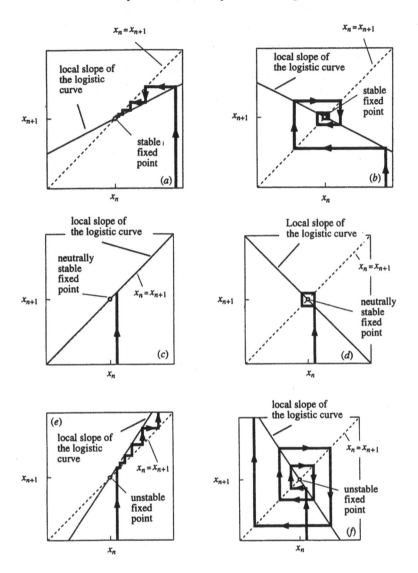

Figure 5.5. The local stability of fixed points. (*a*) Stepping into a stable fixed point. (*b*) Spiralling into a stable fixed point. (*c*) A neutrally stable fixed point (slope = 1). (*d*) A neutrally stable fixed point (slope = −1). (*e*) Stepping out from an unstable fixed point. (*f*) Spiralling out from an unstable fixed point.

This is illustrated graphically in figures 5.5(*a*)–(*f*), where we zoom into the crossing point of the logistic curve and the 45° reflecting line (of slope +1). Although parabolic, the logistic curve may be approximated by a straight line in the immediate vicinity of the crossing point. The fixed points attract the orbit for $\left|f'(x_n)\right| < 1$, neither attract nor repel for $\left|f'(x_n)\right| = 1$ and repel for $\left|f'(x_n)\right| > 1$. The graphical iteration of the orbit is shown as a thick solid line in the figures, with the fixed points marked by small circles. The orbits may be attracted to, or repelled by, the fixed point in two distinct fashions: either by stepping (figures 5.5(a) and (e)) for locally positive slopes of the logistic curve, or spiralling (figures 5.5(b) and (d)) for locally negative slopes of the logistic curve. The behaviour of neutrally stable fixed points is depicted in figures 5.5(c) and (d). With a positive slope (equal to $+1$) of the logistic curve in the vicinity of the fixed point, the iterated solution remains at one point, neither being attracted to, nor repelled by, the fixed point. For a negative slope (equal to -1) of the logistic curve in the vicinity of the fixed point the iterated solution alternates either side of the fixed point, again moving neither away from, or towards, the fixed point. Note that this last case is not to be confused with a period 2 solution, as here we are concentrating on the 'immediate' vicinity of the fixed point and have approximated the logistic parabola as a straight line.

The slope of the logistic curve is

$$f'(x_n) = a(1 - 2x_n) \tag{5.10}$$

and for values of $a < 1$, the logistic map has only one fixed point at the origin as it only crosses the $x_n = x_{n+1}$ line at that point. From equation (5.10) we see that these fixed points of $x_n = 0$ are stable for $a < 1$. When $a > 1$, the fixed point $x = 0$ becomes unstable and a new stable fixed point emerges for the logistic map at $x = (a - 1)/a$ (as we saw in the previous section). Substituting this fixed point into equation (5.10) we arrive at,

$$f'(x_n) = a\left(1 - 2\left(\frac{a-1}{a}\right)\right) = 2 - a. \tag{5.11}$$

Thus, the fixed point at $(a - 1)/a$ is stable up to values of $a < 3$, since this results in an absolute value of $f'(x_n) < 1$. At $a = 1$ and $a = 3$ the fixed point is neutrally stable. Above $a = 3$, the fixed point becomes unstable and a period 2 orbit emerges, where the consecutive iterations alternate between values of x either side of the now unstable original fixed point. The values of the period 2 orbit may be found in the same manner as the $(1 - a)/a$ fixed point. This time, however, we require

$$x_{n+2} = x_n \tag{5.12a}$$

thus the logistic map requires to be iterated twice before returning to the fixed point, i.e.

$$x_{n+2} = ax_{n+1}(1 - x_{n+1}) = a^2 x_n(1 - x_n)(1 - (ax_n(1 - x_n))) \tag{5.12b}$$

equating equation (5.12a) with (5.12b), we find x_n from the fourth-order equation,

$$x_n = a^2 x_n(1 - x_n)(1 - (ax_n(1 - x_n))). \tag{5.12c}$$

Two solutions are already known, namely $x_n = 0$ and $x_n = (a - 1)/a$: these correspond to the two unstable fixed points of the system. The remaining two solutions are

$$x_n = \frac{a + 1 + \sqrt{a^2 - 2a - 3}}{2a} \tag{5.13a}$$

$$x_n = \frac{a + 1 - \sqrt{a^2 - 2a - 3}}{2a}. \tag{5.13b}$$

These are the two stable fixed points of the system described by equation (5.12c), which in turn are the two alternate values of x visited on the period 2 attractor. Note that these solutions are defined only for values of a greater than or equal to 3, where the first bifurcation takes place for the logistic map. At $a = 3$ the initial fixed point of $(1 - a)/a$ loses stability and becomes a repelling fixed point. The two fixed points given by (5.13a, b) are attracting. This is the mechanism by which period doubling of the attractor takes place. At each bifurcation, each fixed point loses its stability and is replaced by two new attracting fixed points (refer back to figure 5.4(c)). To find the period 4, period 8, etc solutions, we respectively set $x_n = x_{n+4}$, $x_n = x_{n+8}$, etc and proceed as above. It is easily seen that the calculation of these roots quickly becomes very involved, even for relatively low periods.

The location of the bifurcation points on the bifurcation diagram is also of interest. First we denote the value of a for the kth bifurcation by a_k. The first bifurcation, from period 1 to period 2, occurs at $a_1 = 3.0$, the second bifurcation, from period 2 to period 4, occurs at $a_2 = 3.449\,490\ldots$ and so on. After each bifurcation point the period is then 2^k. A list of the first seven bifurcation points reads as follows:

$a_1 = 3.0$
$a_2 = 3.449\,490\ldots$
$a_3 = 3.544\,090\ldots$
$a_4 = 3.564\,407\ldots$
$a_5 = 3.568\,759\ldots$
$a_6 = 3.569\,692\ldots$
$a_7 = 3.569\,891\ldots.$

The ratios of successive separations of the bifurcation points are given by,

$$\frac{a_2 - a_1}{a_3 - a_2} \approx \frac{a_3 - a_2}{a_4 - a_3} \approx \frac{a_4 - a_3}{a_5 - a_4} \approx \frac{a_5 - a_4}{a_6 - a_5} \approx \frac{a_6 - a_5}{a_7 - a_6} \approx \ldots \approx \frac{a_k - a_{k-1}}{a_{k+1} - a_k} \quad (5.14)$$

and have a particular scaling associated with them. The ratios tend to a constant as k tends to infinity: more formally

$$\lim_{k \to \infty} \left[\frac{a_k - a_{k-1}}{a_{k+1} - a_k} \right] = \delta = 4.669201\ldots. \quad (5.15)$$

The ratio δ is known as the **Feigenbaum number**. The Feigenbaum number was first found in the iteration of the logistic map, however, it has now been found in many other systems, both mathematical and, more importantly for us, real systems. δ is a universal constant, just like π or e.

Many 1D maps have now been investigated which contain period doubling bifurcations in which δ may be found. Some of the more common maps are listed below.

(i) The sine map.

$$x_{n+1} = ax^2 \sin(\pi x_n). \quad (5.16)$$

(ii) The tent map.

$$x_{n+1} = r(1 - |2x_n - 1|). \quad (5.17)$$

(iii) The Baker map.

$$x_{n+1} = 2x_n \mod 1. \tag{5.18}$$

(iv) The circle map.

$$x_{n+1} = x_n + r - \frac{K}{2\pi} \sin(2\pi x_n). \tag{5.19}$$

(v) The Gaussian map.

$$x_{n+1} = e^{-bx_n^2} + c. \tag{5.20}$$

Iterating the above maps for various values of the control parameters produces a rich variety of behaviours. (Why not try iterating them for yourself?) References for further maps are to be found at the end of this chapter.

5.8   A two-dimensional map: the Hénon model

The Hénon model, or **Hénon map** (equations (5.21a, b)) is interesting in that it consists of two variables x and y. Both chaotic and periodic attractors may be found for the Hénon model depending on the choice of the control parameters a and b.

$$x_{n+1} = 1 - ax_n^2 + y_n \tag{5.21a}$$

$$y_{n+1} = bx_n. \tag{5.21b}$$

In table 5.1, the Hénon map is iterated for two sets of the control parameters. The first set ($a = 0.9$, $b = 0.3$) leads to a post-transient orbit on a periodic attractor. The second set of control parameters ($a = 1.4$, $b = 0.3$) leads to a chaotic sequence of iterates. The table contains the first ten iterates of the Hénon model for both parameter sets together with iterate numbers 250–256, by which time the transients have decayed enough for us to see the periodic solution for the ($a = 0.9$, $b = 0.3$) case. The chaotic case never repeats and is better visualized by plotting out the iterations.

Figures 5.6(a) and (b) contain plots of the iterated solutions to the Hénon map for the two sets of control parameters, both iterations begin at ($x_0 = 0.8$, $y_0 = 0.8$). In figure 5.6(a), time series of both x and y are plotted together with a plot of the iterates in the x–y plane. From the time series plots of the first 100 iterates, the convergence of the transient orbits onto a period 2 attractor is evident. In the x–y plot the initial condition and the first two mapped points are connected by arrows. The initial point (0.8, 0.8) is mapped onto (1.224, 0.240), this point is then mapped onto ($-0.1083...$, 0.3672). As the iterations proceed, the transient points converge onto the period 2 attractor with co-ordinates ($-0.4262...$, 0.3606) and (1.2302..., -0.1251). In figure 5.6(b), the same initial condition of (0.8, 0.8) is used. This time however, the mapped points are chaotic and quickly converge onto the strange attractor which has the general shape of a braided, back-to-front letter 'C'. In fact, the convergence is so rapid that only the first two transient values are obvious in the x–y plot, namely the initial point

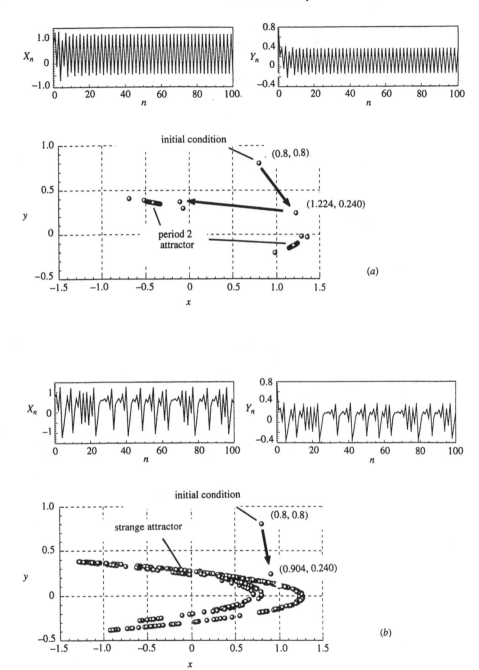

Figure 5.6. The Hénon attractor. (*a*) The period 2 attractor for the Hénon model with $a = 0.9$, $b = 0.3$. (*b*) The strange attractor for the Hénon model with $a = 1.4$, $b = 0.3$. (*c*) The whole attractor ($a = 1.4$, $b = 0.3$). (*d*) Zoom 1. (*e*) Zoom 2. (Continued overleaf.)

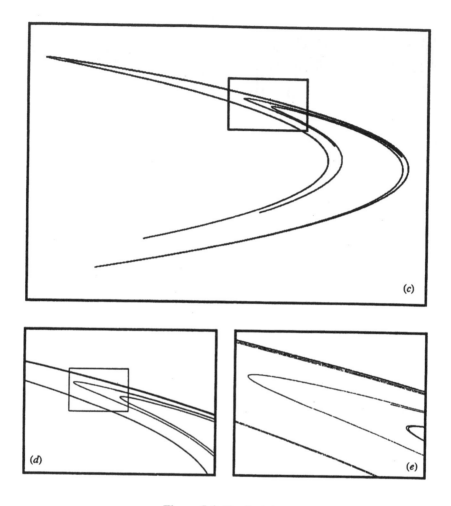

Figure 5.6. Continued.

(0.8, 0.8) and the first iterated point (0.904, 0.240). Although the time series appears to consist of many similar structures, none of these are identical as the orbit never repeats.

If we plot enough post-transient iterates of the Hénon map we may visualize the fine structure of its strange attractor. This is done in figure 5.6(c), which is the strange attractor of figure 5.6(b) plotted using many more points. Figure 5.6(d) shows a blow up of the boxed part of the Hénon attractor in figure 5.6(c). Figure 5.6(e) shows a blow up of the boxed part of figure 5.6(d). By figure 5.6(e) we see that what appeared as solid lines in the previous figures have now split up into bands. As we zoom into the bands of the Hénon attractor more and more structure is revealed. The strange attractor for the Hénon map has a **fractal structure**, a key property of a strange attractor. Moreover, the attractor has a Cantor-set-like construction across the bands. Using the box counting method described in chapter 3, the fractal dimension of the Hénon attractor has been found to be around 1.28.

Table 5.1. The iterated solutions to the Hénon map.

	Control parameter settings			
	$a = 0.9, b = 0.3$		$a = 1.4, b = 0.3$	
n	x_n	y_n	x_n	y_n
0	0.8	0.8	0.8	0.8
1	1.224	0.24	0.904	0.24
2	$-0.1083\ldots$	0.3672	$0.0958\ldots$	0.2712
3	$1.3566\ldots$	$-0.0325\ldots$	$1.2583\ldots$	$0.0287\ldots$
4	$-0.6889\ldots$	$0.4069\ldots$	$-1.1879\ldots$	$0.3774\ldots$
5	$0.9798\ldots$	$-0.2066\ldots$	$-0.5982\ldots$	$-0.3563\ldots$
6	$-0.0707\ldots$	$0.2939\ldots$	$0.1425\ldots$	$-0.1794\ldots$
7	$1.2894\ldots$	$-0.0212\ldots$	$0.7920\ldots$	$0.0427\ldots$
8	$-0.5176\ldots$	$0.3868\ldots$	$0.1643\ldots$	$0.2376\ldots$
9	$1.1456\ldots$	$-0.1552\ldots$	$1.1997\ldots$	$0.0493\ldots$
10	$-0.3366\ldots$	$0.3437\ldots$	$-0.9659\ldots$	$0.3599\ldots$
.	.	.	.	.
.	.	.	.	.
.	.	.	.	.
250	$-0.4262\ldots$	$0.3606\ldots$	$-1.2807\ldots$	$0.3808\ldots$
251	$1.2302\ldots$	$-0.1251\ldots$	$-0.9154\ldots$	$-0.3842\ldots$
252	$-0.4262\ldots$	$0.3606\ldots$	$-0.5575\ldots$	$-0.2746\ldots$
253	$1.2302\ldots$	$-0.1251\ldots$	$0.2901\ldots$	$-0.1672\ldots$
254	$-0.4262\ldots$	$0.3606\ldots$	$0.7148\ldots$	$0.0870\ldots$
255	$1.2302\ldots$	$-0.1251\ldots$	$0.3716\ldots$	$0.2144\ldots$

Table 5.2. The first ten iterated solutions to the complex map for various initial conditions and control parameter values.

	Control parameter settings					
	$c = -1 + 0i$		$c = -1 + 0i$		$c = -1.1 + 0.2i$	
n	Real z_n	Imag. z_n	Real z_n	Imag. z_n	Real z_n	Imag. z_n
0	0	0	0	1	0	0
1	-1	0	-2	0	-1.1	0.2
2	0	0	3	0	0.07	0.24
3	-1	0	8	0	-1.1527	0.1664
4	0	0	63	0	$0.2010\ldots$	$0.1836\ldots$
5	-1	0	$3\,968$	0	$-1.0933\ldots$	$0.1261\ldots$
6	0	0	$15\,745\,023$	0	$0.0793\ldots$	$-0.0758\ldots$
7	-1	0	2.479×10^{14}	0	$-1.0994\ldots$	$0.1879\ldots$
8	0	0	6.145×10^{28}	0	$0.0734\ldots$	$-0.2132\ldots$
9	-1	0	3.77×10^{57}	0	$-1.1400\ldots$	$0.1686\ldots$
10	0	0	1.42×10^{115}	0	$0.1713\ldots$	$-0.1845\ldots$

5.9   Iterations in the complex plane: Julia sets and the Mandelbrot set

This section contains a brief overview of iterations of the complex map,

$$z_{n+1} = z_n^2 + c \qquad (5.22a)$$

where z is a complex number with the general form,

$$z = a + bi \qquad (5.22b)$$

where a and b are real numbers, and

$$i = \sqrt{-1}. \qquad (5.22c)$$

As with the logistic map, the iterated solutions to the complex map of equation (5.22a) may exhibit a variety of behaviours dependent upon both the control parameter, c, and the initial condition, z_0, used.

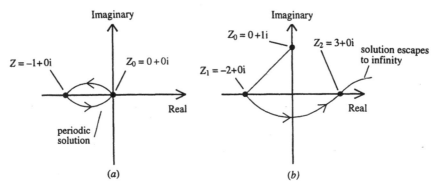

Figure 5.7. Iterations of the complex map with $c = (-1 + 0i)$. (*a*) Iteration begins at $(0, 0i)$. (*b*) Iteration begins at $(0, i)$.

First, we will investigate the effect of the initial condition on the sequence of iterated solutions (orbit) to the complex map by keeping c constant at $(-1+0i)$ and iterating the map using the following initial values of z: $z_0 = (0+0i)$ and $z_0 = (0+1i)$. Columns 2–5 of table 5.2 list the iterations and figure 5.7 shows schematically the iteration process. The iteration begun at $z_0 = (0 + 0i)$ produces a periodic orbit which oscillates between $(0 + 0i)$ and $(-1 + 0i)$. However, the iteration begun at $z_0 = (0 + 1i)$ increases in magnitude, escaping to infinity as the number of iterations go to infinity. If we continue to iterate the map for other values of z_0 we will find that some iterations escape to infinity and others are attracted to a period 2 orbit. All the initial values of z which do not escape to infinity are plotted in the complex z plane in figure 5.8(*a*). This set of points, shown shaded in the figure, is known as the **prisoner set** of the complex function. The prisoner set is the basin of attraction for the periodic attractor. The set of all points on the boundary between the prisoner set and all the points which escape to infinity is known as the **Julia set**. The Julia set in figure 5.8(*a*) is **connected**, that is, all points in the set can be reached from all other points in the set along some path without leaving the set.

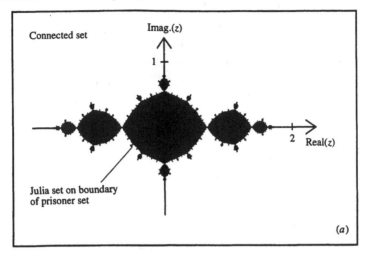

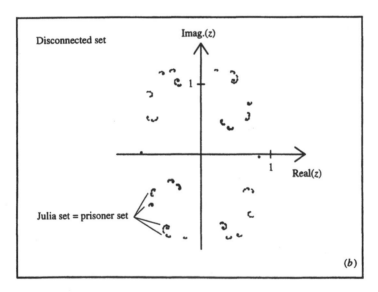

Figure 5.8. Prisoner sets of the complex map. (*a*) $c = (-1 + 0i)$. (*b*) $c = (0.5 + 0.1i)$.

However, for some values of the control parameter c the Julia set is **disconnected**, see for example figure 5.8(*b*), which shows the Julia set for $c = (0.5 + 0.1i)$. Disconnected Julia sets have a Cantor-set-like structure in the plane.

The values of $z_0 = (0, 0i)$ and $c = (-1 + 0i)$ used in the above illustration gave orbits for the complex map which actually began on the attractor of $[(0, 0i), (-1, 0i)]$, and so no transient was produced. The last two columns of table 5.2 contain the first ten iterations of the map for $c = (-1.1, 0.2i)$ with $z_0 = (0, 0i)$, no obvious behaviour of the orbit is evident from the table. Figure 5.9(*a*) shows the first sixty iterations of the time

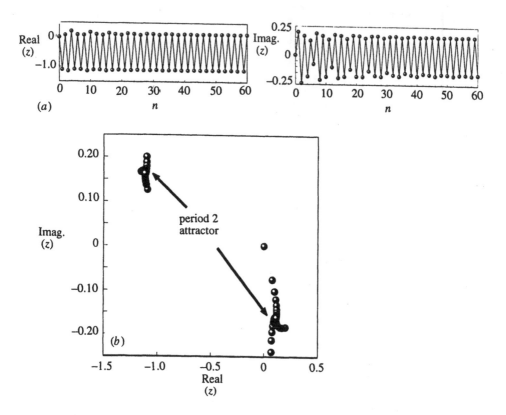

Figure 5.9. Iterations of the complex map with $c = (-1.1 + 0.2i)$. (*a*) The time series behaviour of the real and imaginary parts of the iterations of the complex map. (*b*) Transient iterations of the complex map in the complex plane. The orbits are heading towards a period 2 attractor at points $(-1.1138, 0.1631)$ and $(0.1131, -0.1631)$.

series of both the real and imaginary parts of z. Figure 5.9(*b*) plots these iterations in the complex plane. From both figures, we see the orbit approaching a period 2 attractor asymptotically as the number of iterations, n, goes to infinity. This is the same type of transient behaviour which we found in the logistic map and the Hénon attractor.

To iterate enough maps to visualize the resulting Julia set takes a lot of computational effort. However, it is a much simpler task to determine whether a particular value of the control parameter c produces a connected or disconnected Julia set. To do this, we only require to iterate the complex map, for each c value, with the initial value of z set to $z_0 = (0 + 0i)$. If this point escapes to infinity then the Julia set corresponding to that particular value of c is disconnected, otherwise the set is connected. Connected Julia sets correspond to periodic or chaotic orbits which do not escape.

Now, if we plot all the values of c which produce connected Julia sets on the complex c-plane, we end up with the **Mandelbrot set**, shown as the black region of

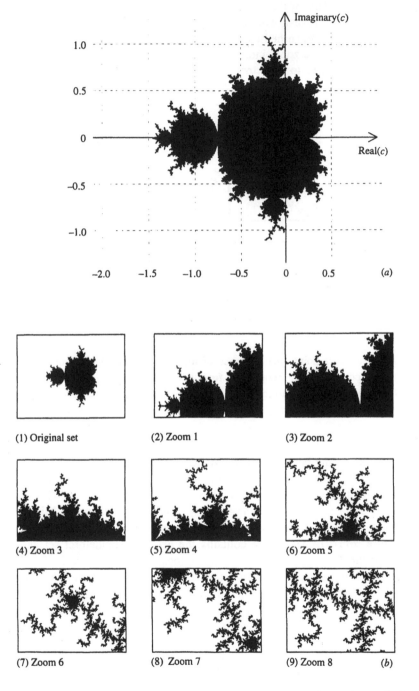

(1) Original set   (2) Zoom 1   (3) Zoom 2

(4) Zoom 3   (5) Zoom 4   (6) Zoom 5

(7) Zoom 6   (8) Zoom 7   (9) Zoom 8   (b)

Figure 5.10. The Mandelbrot set. (a) The Mandelbrot set in the complex c-plane. (b) Zooming into the Mandelbrot set.

figure 5.10(*a*). The boundary of the Mandelbrot set has an infinitely complex structure. Zooming into the boundary reveals ever more detail. This is done in figure 5.10(*b*) where each successive image is a magnified part of the previous image. See if you can follow the sequence of images as we zoom into the boundary of the Mandelbrot set. Different regions of the set correspond to attractors for the complex map of various periods. A period doubling route to chaos may be observed for the control parameter c along the real axis from right to left.

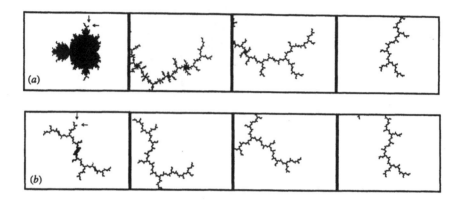

Figure 5.11. A Misiurewicz point. (*a*) Sequence of images from the Mandelbrot set, zooming into the Misiurewicz point at $(0, i)$. Zooming in from left to right to the point $(0, i)$, located by arrows in the first image. (*b*) Sequence of images from the Julia set for $c = (0, i)$, zooming into the Misiurewicz point at $(0, i)$. Zooming in from left to right to the point $(0, i)$, located by arrows in the first image.

At certain points on the boundary of the Mandelbrot set, the Mandelbrot set and the Julia set, corresponding to a particular c value, look similar and, as the magnification is increased, the sets begin to appear identical. These points, where the Mandelbrot set and the corresponding Julia set become asymptotically similar, are known as **Misiurewicz points**. The Mandelbrot set and corresponding Julia set at one such Misiurewicz point is shown in figure 5.11(*a*). The figure contains the original Mandelbrot set and three zoomed images at the Misiurewicz point $c = (0, i)$, located by arrows in the first image. Figure 5.11(*b*) contains the corresponding Julia set for $z = c = (0, i)$, and three zoomed images of it at the point $(0, i)$, again located by arrows in the first image. The similarity between the last images of figures 5.11(*a*) and (*b*) is striking, although a slight rotation is evident. In fact, as we zoom into Misiurewicz points, the Mandelbrot and Julia sets become increasingly similar at the point c under scaling and rotation.

In addition to Misiurewicz points, other points within the boundary of the Mandelbrot set contain smaller Mandelbrot-set-like-structures, two such mini-Mandelbrot set structures are shown in figure 5.12, where we zoom into one of the top filaments of the Mandelbrot set. The locations of the zoomed images are shown boxed in the previous image. In fact these mini-Mandelbrot sets abound and are to be found at all scales within the boundary of the Mandelbrot set. The Mandelbrot set has been called

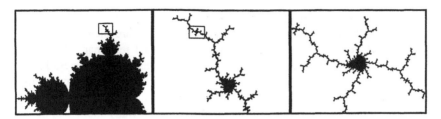

Figure 5.12. Mini-Mandelbrot sets. Sequence of images from the Mandelbrot set, zooming into one of the branches to reveal smaller Mandelbrot set structures. Zooming in from left to right.

a 'borderline fractal' as it contains many similar structures along its boundary, however, exact self-similarity is never achieved.

5.10 Chapter summary and further reading

5.10.1 Chapter keywords and key phrases

discrete dynamical systems	*Verhulst model*	*nonlinear feedback*
logistic map	*non-invertible*	*orbit*
control parameter	*logistic curve*	*time series*
periodic orbit	*chaotic orbit/chaos*	*basin of attraction*
periodic attractor	*chaotic/strange attractor*	*bifurcation diagram*
period doubling bifurcation	*period doubling route*	*Mandelbrot set*
Feigenbaum number	*Hénon map*	*Misiurewicz points*
fractal structure (of strange	*prisoner set*	*connected/disconnected set*
attractor)	*Julia set*	*attractor*

5.10.2 General

In this chapter we have seen how the iteration of relatively simple maps can lead to very complex dynamics. Using a control parameter, we can change the output of the nonlinear logistic map from periodic orbits to chaotic orbits via a universal period doubling cascade characterized by the Feigenbaum number (Feigenbaum 1978, 1980). The resulting bifurcation diagram contains infinite detail, with chaotic and periodic attractors intermixed. The reader is referred to the article by Metropolis *et al* (1973) for more information on the universal sequence of periodic attractors in unimodal maps and the application of symbolic dynamics (not covered herein) in the detection of the sequence.

5.10.3 Maps

We have concentrated on the 1D logistic map (May 1974, 1976), the 2D Hénon map (Hénon 1976) and the complex map associated with the Mandelbrot set (Mandelbrot 1980, 1983). For more information on the properties of the maps and their associated bifurcation diagrams see for example the texts by Froyland (1992) or Peitgen *et al* (1992a); see also the collection of articles by Devaney and Keen (1989). A more elementary text by Peitgen *et al* (1992b) deals in detail with the logistic map and tent map. The text by Becker and Dörfler (1989) provides many computer programs for the generation of a variety of maps, both real and complex. The mathematically inclined reader is referred to the text by Holmgren (1994), which gives a broad introductory coverage of discrete dynamical systems including the logistic and complex maps. Many more examples of maps are to be found in the literature. The reader is referred to the books by Barnsley and Demko (1986), Froyland (1992) and Korsch and Jodl (1994) for a selection of 1D and 2D maps and their properties. See the work by Hilborn (1994) for more details of the properties of the **Gaussian map**. Borcherds and McCauley (1993) discuss the **digital tent map** and the **trapezoidal map**. Li and Yorke (1975) show how the existence of a period 3 attractor implies that the system will display chaos. Numerical methods to find the roots to polynomials, such as **Newton's method**, approach the roots using an iterative procedure. The root approached depends upon the initial conditions chosen in the iteration. The **basin of attractions** resulting from these methods may exhibit very rich structures: see for example the articles by Walter (1994), Gilbert (1994) or Qammar and Mossayebi (1996).

Many variations of the logistic map have been investigated by various researchers. The variants include those driven by, or containing, **noise** (Malescio 1996, Choi and Lee 1995, Gutierrez *et al* 1993), maps with **memory** (Hartwich and Fick 1993), **coupled map systems** (Gade and Amritkar 1994), maps with **delay** (Cabrera and de la Rubia 1995), and maps with a **linear** part (Tan and Chia 1995). See also the article by Phatak and Rao (1995), who consider the logistic map as a possible **random number generator**. Pressing (1988) discusses the generation of **music** using a variety of nonlinear maps including the logistic map and the complex map, in the same vein see Mayer-Kress *et al* (1992). Thiran and Setti (1995) discuss the use of 1D maps for **storing and retrieving information** and Erramilli *et al* (1995) investigate the application of chaotic maps to model **telecommunication traffic**. May (1976) discusses the logistic map in the context of **biological populations**. Grassberger (1981) discusses the **fractal dimension** of the chaotic attractors for the logistic equation. Later, Grassberger (1983a) discusses the computation of the **fractal dimension** estimate for the Hénon attractor. The effect of the control parameters a and b on the behaviour of the Hénon map is investigated by Gallas (1993). Pando *et al* (1995) show that a CO_2 **laser** may behave in a qualitatively similar manner to a Hénon map.

5.10.4 The Mandelbrot set

It is impossible to do justice to the infinite wonder of the Mandelbrot set within the limited space allowed here. Many authors have spent a great deal of (computational) time and effort in illustrating the infinite complexity of the Mandelbrot set and associated Julia sets. An excellent place to start is with the book by Peitgen and Richter (1986) which contains many detailed colour images of the Mandelbrot set embedded within an enlightening text (it also includes many fine pictures of basins of attraction from Newton's method). A collection of articles edited by Pickover (1995) is also worth consulting for many detailed images of the Mandelbrot set, Julia sets and other mappings. Even better, the video by

Peitgen *et al* (1990) provides the viewer with some stunning 'flight' sequences over and into the Mandelbrot set. A rigorous treatment of the self-similarity of Misiurewicz points in the Mandelbrot and corresponding Julia sets is given by Lei (1990), together with the similarity between the Mandelbrot and Julia sets at such a point. Douady and Hubbard (1985) explain the relationship of mini-Mandelbrot sets to the whole set. The linear and angular distortions of the mini-Mandelbrot sets are discussed by Philip *et al* (1994). Romera *et al* (1996) provide some graphical tools to investigate the Mandelbrot set. For further illustrations of the properties of the Mandelbrot and Julia sets the reader is referred to the papers by Cvitanovic and Myrheim (1989), Bhavsar *et al* (1993), Philip (1992), Frame *et al* (1992), Lutzky (1993), Pickover (1994), Balkin *et al* (1994), Miller (1993), Sherard (1993), Shishikura (1994) and Dixon *et al* (1996). As with the logistic map, there are numerous variations of the complex map to be found in the literature. Generalizations of the complex map and associated Mandelbrot set are discussed by Frame and Robertson (1992); Kawabe and Kondo (1993) discuss the properties of a logarithmic version of the complex map; Metzler (1994) discusses the filled-in square structure of the quadratic Mandelbrot set; and Gujar *et al* (1992) and Dhurandhar *et al* (1993) consider the complex map with generalized real number exponents. Two-dimensional quadratic iterated maps (of which the complex map discussed in this chapter is a specific example) are discussed by Sprott and Pickover (1995), and Sprott (1993) investigates the strange attractors associated with 2D and 3D quadratic maps.

5.10.5 Iterated function systems

Finally, the reader is pointed towards the detailed text on iterated function systems, IFSs, by Barnsley (1988). IFSs are outside the scope of this text: they are, essentially, sets of rules with which whole images are iterated towards final attractors consisting of complex (and beautiful) fractal shapes.

5.11 Revision questions and further tasks

(Note: If you have access to a spreadsheet software package, you may find it very useful in speeding up the iteration process in some of the questions below, otherwise, a hand held calculator will suffice.)

Q5.1 List the keywords and key phrases of this chapter and briefly jot down your understanding of them.

Q5.2 Find the post-transient solution to the logistic map, $x_{n+1} = ax_n(1 - x_n)$, using the following values of the control parameter: $a = 0.9, 2.8, 3.18, 3.5, 4$. Iterate the map twenty times for each value of a (more if required) using $x_0 = 0.2$. Tabulate the results to four significant figures and suggest what the period of the post-transient orbit might be in each case.

Q5.3 Using the data obtained in Q5.2, plot out the time series (i.e. x_n against n) of the orbits of the maps for each of the five values of the control parameter, a.

Q5.4 Use figure Q5.4 to graphically iterate the logistic map for the control parameter set to $a = 1.9$ and $a = 2.18$. Begin the iteration at the initial condition indicated by the

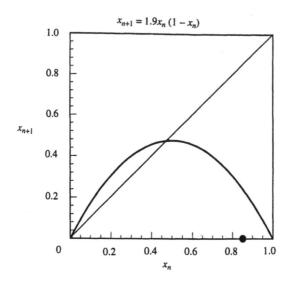

$$x_{n+1} = 1.9x_n(1 - x_n)$$

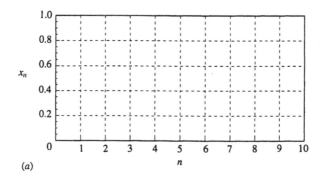

(a)

Figure Q5.4. Graphical iteration of the logistic equation. (*a*) *a* = 1.9. (*b*) *a* = 3.18. (Continued overleaf.)

dot on the x_n axis. Plot out the time series on the lower graphs provided for the first ten iterates.

Q5.5 Locate the post-transient solutions to the logistic map that you obtained in Q5.2 on the bifurcation plots in figure 5.4. Explain why a computer is a necessity to produce such a diagram with sufficient resolution to visualize the behaviour of the system.

Q5.6 The bifurcation points, a_k, of the logistic map are listed as follows: $a_1 = 3.000\,000\ldots$; $a_2 = 3.449\,490\ldots$; $a_3 = 3.544\,090\ldots$; $a_4 = 3.564\,407\ldots$; $a_5 = 3.568\,759\ldots$; $a_6 = 3.569\,692\ldots$; $a_7 = 3.569\,891\ldots$. Use these numbers to show that the Feigenbaum number tends towards its universal constant value of $4.669\,202\ldots$, as the bifurcation number, k, increases.

$$x_{n+1} = 3.18x_n (1- x_n)$$

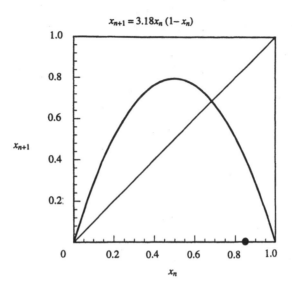

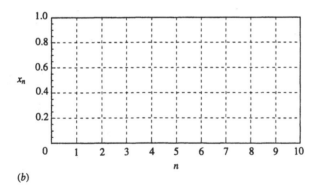

(b)

Figure Q5.4. Continued.

Q5.7 Repeat Q5.2 for the logistic map with control parameter, a, set at 2.8 and 4. This time start the process at initial values of $x = 0.5$ and 0.8. What can you conclude from the results?

Q5.8 Describe the attractor for the sine mapping $x_{n+1} = ax^2 \sin(\pi x_n)$ for values of the control parameter, $a = 1.8$ and 2.1. Iterate the mapping twenty times using initial values of x of 0.5, 0.75, and 0.9. (Note that x is in radians.)

Q5.9 Plot the first twenty points of the Hénon attractor in x–y space. Use control parameters $a = 1.4$; $b = 0.3$ and initial values $x = 0.0$; $y = 0.0$. Try to locate a set of control parameters, a and b, which give a periodic Hénon attractor.

Q5.10 Calculate the first five iterations of the complex map $z_{n+1} = z_n^2 + c$ using a value of $c = -1+0i$, and initial values of z of $(-1+0i)$ and $(1.5+1.5i)$ respectively. Plot the iterates in the complex plane. Which of the two initial z values belongs to the prisoner set?

Q5.11 What is the boundary of a prisoner set known as?

Q5.12 Determine whether each of the following complex maps produce connected or disconnected Julia sets:

$$z_{n+1} = z_n^2 - 1 \qquad z_{n+1} = z_n^2 + 0.5 + 0.1i \qquad z_{n+1} = z_n^2 + 2 - 0.4i \qquad z_{n+1} = z_n^2 - 0.5 - 0.2i.$$

Q5.13 Locate the positions of the complex functions of Q5.12 on the c plane of figure 10(a) in the text. Explain the fundamental difference between those points contained within the Mandelbrot set and those which exist outside the Mandelbrot set.

Q5.14 Using only the information gained in Q5.12 and Q5.13, determine which of the following complex maps produce disconnected Julia sets:

$$z_{n+1} = z_n^2 + 0.5 - 0.1i \qquad z_{n+1} = z_n^2 + 2 + 0.4i \qquad z_{n+1} = z_n^2 - 0.5 + 0.2i.$$

Q5.15 Determine the periodicity of the solutions to the following complex maps. Iterate the maps for initial values of z of $(0 + 0i)$. What do you think is happening to the attractor?

$$z_{n+1} = z_n^2 \quad z_{n+1} = z_n^2 - 0.5 \quad z_{n+1} = z_n^2 - 1.0 \quad z_{n+1} = z_n^2 - 1.35 \quad z_{n+1} = z_n^2 - 1.37.$$

Chapter 6

Chaotic oscillations

6.1 Introduction

In this chapter we concentrate on the oscillatory behaviour of relatively simple nonlinear dynamical systems which evolve continuously in time, and whose temporal evolutions are defined by differential equations of motion. As with their discrete counterparts, discussed in chapter 5, continuous-time nonlinear systems may exhibit either periodic or chaotic oscillations. (Linear systems cannot exhibit chaos.) Towards the end of the chapter we will briefly look at more complex systems which can exhibit both spatial and temporal complexity.

6.2 A simple nonlinear mechanical oscillator: the Duffing oscillator

Figure 6.1(a) provides a sketch of a simple sliding block oscillator model consisting of a mass, m, attached to a wall by a spring with stiffness s, a frictional resistance r between the block and the sliding surface and a force F acting on the mass. This model can produce chaos under the right conditions. We shall 'build' such a chaos producing oscillator from the elements given in the sketch. It will also help remind ourselves of simple oscillator theory.

Let us begin by setting the frictional resistance and forcing to zero. If we move the mass from its rest position we would expect to feel a force, caused by the stretched spring, which wants to pull the mass back to the unstretched, or rest, position. This restoring force, F_S, is simply equal to the spring stiffness multiplied by the extension of the spring, i.e. $F_S = -sx$. (The negative sign indicating that the restoring force is in the opposite direction to the extension, as the spring acts to resist the extension). If we pull the mass to a new position, a distance A from its rest position (a stable equilibrium at $x = 0$), and let go, the mass will oscillate back and forth between $+A$ and $-A$ from the rest position. The time series of this undamped simple harmonic motion is given in figure 6.1(b), and the period of oscillation T will be equal to $2\pi\sqrt{m/s}$. If we include the frictional resistance of the block sliding on the surface by adding a frictional or damping component r and once again release the mass from A, the mass will oscillate back and forth, this time with ever decreasing amplitude. As with the spring force, the frictional resistance force F_r acts against the motion and in this discussion we assume it

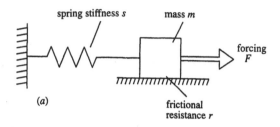

(a)

spring stiffness s mass m

forcing F

frictional resistance r

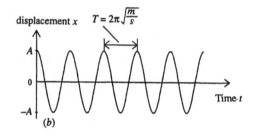

displacement x $T = 2\pi\sqrt{\frac{m}{s}}$

A

0

$-A$

Time t

(b)

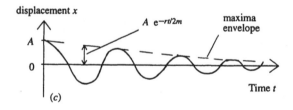

displacement x

$A\,e^{-rt/2m}$ maxima envelope

A

0

Time t

(c)

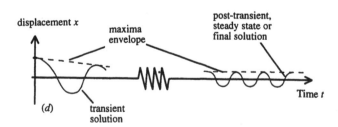

displacement x

maxima envelope

post-transient, steady state or final solution

Time t

(d)   transient solution

Figure 6.1. Sliding block oscillator model. (a) The forced, damped oscillator model. (b) The time series of the simple harmonic oscillations. (c) The lightly damped simple harmonic motion of the unforced, damped linear oscillator model. (d) Typical transient and post-transient solutions.

to be directly proportional to the velocity of the block. The amplitude of the oscillations decays exponentially, damped down by the friction in the system. A typical time series of this damped simple harmonic motion is given in figure 6.1(*c*). If we now apply a time varying forcing to the model system then we may keep the oscillations going indefinitely. The simplest forcing we can apply which will achieve this is a sinusoidal force. If we again begin the oscillation, this time from an arbitrary initial displacement, the time evolution generally consists of two regions: the first, an initial **transient oscillation** which takes the dynamical system from the initial condition to the second, its **post-transient oscillation** (figure 6.1(*d*)). From its initial condition, the dynamical system is attracted to this post-transient solution, or **attractor**. Post-transient attractors are typical of dissipative dynamical systems, in this case the dissipation occurs due to the friction term in the equations of motion. Once on the post-transient solution, the system stays there unless disturbed by an external force. In this text we concentrate exclusively on dissipative dynamical systems.

The post-transient solution is periodic: to obtain a chaotic post-transient solution we need to add one further ingredient to our dynamical system, that is nonlinearity. A simple way to do this is to have a nonlinear spring, where the restoring force of the spring is proportional to the cube of the displacement, i.e. $F_S = -sx^3$. This simple forced, damped nonlinear oscillator is known as the **Duffing oscillator**, and has the equation of motion

$$m\ddot{x} + r\dot{x} + sx^3 = A_f \cos \omega t. \tag{6.1}$$

Here we use the compact nomenclature where displacement is x, velocity, the first derivative with respect to time of the displacement, dx/dt, is denoted by $\dot{x}$, and acceleration, the second time derivative of the displacement, d^2x/dt^2, is denoted by $\ddot{x}$. The dynamical system described by the second-order ordinary differential equation (ODE) of equation (6.1) may exhibit periodic or non-periodic (chaotic) motion depending upon the parameter values chosen for the system. We should note that analytical solutions to the equations of motion of nonlinear systems are generally not possible: this is certainly true for the interesting chaotic cases. We may, however, generate numerical solutions to nonlinear equations using numerical methods such as the Runge–Kutta, the Newmark or the finite difference method. These methods are fairly involved and are outside the scope of this course: see the further reading section at the end of this chapter for more information. All the chaotic behaviours detailed in the examples in this chapter have been generated using the fourth-order Runge–Kutta method (given in the computer program in appendix A).

Without loss of generality, we may simplify the equation of motion of the Duffing oscillator (equation (6.1)) by setting the mass, m, spring stiffness, s, and the angular frequency, ω, to unity, to get

$$\ddot{x} + r\dot{x} + x^3 = A_f \cos t \tag{6.2}$$

We now have only two **control parameters**: a damping coefficient, r, and the amplitude of forcing, A_f. By varying these two parameters we can locate regimes of periodic and chaotic oscillations. For example, periodic behaviour is found for parameter values of $r = 0.08$ and $A_f = 0.20$, while for $r = 0.05$ and $A_f = 7.50$ the behaviour of the system is chaotic.

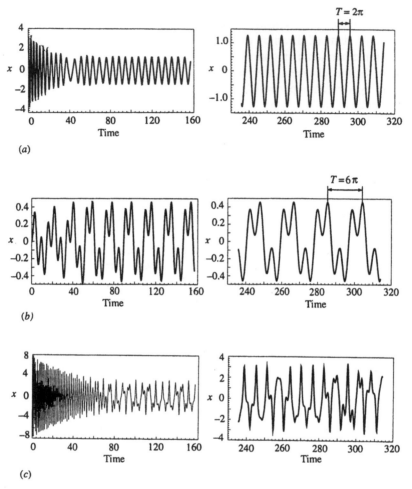

Figure 6.2. Time series for the Duffing oscillator. (*a*) Transient and post-transient time series for the periodic case ($r = 0.08$, $A_f = 0.20$). Initial conditions: $x = 3.0$, $\dot{x} = 3.0$. (*b*) Transient and post-transient time series for the periodic case ($r = 0.08$, $A_f = 0.20$). Initial conditions: $x = 0.0$, $\dot{x} = 0.0$. (*c*) Transient and post-transient time series for the chaotic case ($r = 0.05$, $A_f = 7.5$). Initial conditions: $x = 8.0$, $\dot{x} = 0.0$.

Often, more than one periodic, post-transient solution is possible in such a system, each representing the long-term motion of the oscillator. The post-transient solution picked by the oscillator depends upon the initial conditions of the system, that is the initial displacement, x_0, and velocity $\dot{x}_0$. The set of initial conditions which leads to a particular post-transient solution is known as the **basin of attraction** of the attractor. (See also chapter 5 where we covered the basin of attractions for maps.) In fact, for equation (6.2) with control parameters set to $r = 0.08$ and $A_f = 0.20$, the Duffing oscillator may be attracted towards many post-transient states. Two of these are shown in figure 6.2(*a*) and (*b*), where the transient and post-transient time series are given for both. The Duffing oscillator started at $x_0 = 3.0$, $\dot{x}_0 = 3.0$ ends up on a large amplitude

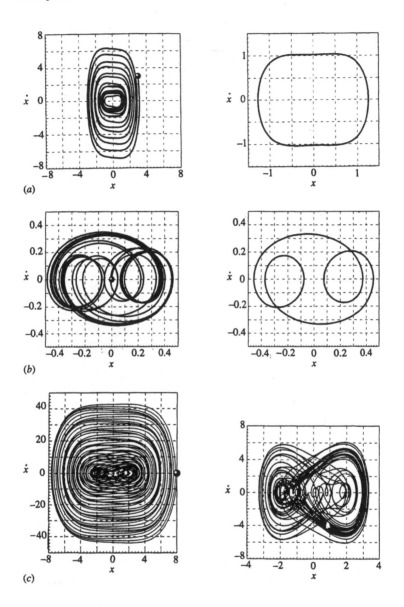

Figure 6.3. Phase portraits of the Duffing oscillator. (*a*) Periodic attractor: transient and post-transient phase portrait for the periodic case. Control parameters: $r = 0.08$, $A_f = 0.20$. Initial conditions: $x = 3.0$, $\dot{x} = 3.0$—marked by the dot in the left-hand part. (*b*) Periodic attractor: transient and post-transient phase portrait for the periodic case. Control parameters: $r = 0.08$, $A_f = 0.20$. Initial conditions: $x = 0.0$, $\dot{x} = 0.0$. (*c*) Strange attractor: transient and post-transient phase portrait for the chaotic case. Control parameters: $r = 0.05$, $A_f = 7.5$. Initial conditions: $x = 8.0$, $\dot{x} = 0.0$. (Note anisotropic scaling of axes.)

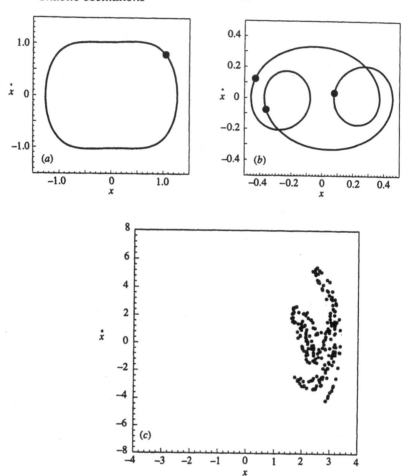

Figure 6.4. Poincaré sections for post-transient solutions of the Duffing oscillator. (*a*) One sampled point for the period 1 oscillation. Control parameters: $r = 0.08$, $A_f = 0.20$. Initial conditions: $x = 3.0$, $\dot{x} = 3.0$. (*b*) Three sampled points for the period 3 oscillation. Control parameters: $r = 0.08$, $A_f = 0.20$. Initial conditions: $x = 0.0$, $\dot{x} = 0.0$. (*c*) Multiple sampled points for the chaotic oscillation. Control parameters: $r = 0.05$, $A_f = 7.5$. Initial conditions: $x = 8.0$, $\dot{x} = 0.0$.

oscillation repeating every period $T = 2\pi$, which is the period of the forcing function. Oscillations which begin at $x = 0.0$, $\dot{x} = 0.0$ settle onto a cycle which contains peaks at intervals of 2π, in this case however, three distinctly different peaks may be observed, and the oscillation now repeats itself every three periods of the forcing function, i.e. at intervals of 6π.

A chaotic oscillation for the Duffing oscillator is shown in figure 6.2(*c*). This corresponds to the control parameters set to $r = 0.05$, $A_f = 7.5$, and an initial condition of $x = 8.0$, $\dot{x} = 0.0$. Once the high-frequency transient disappears, the oscillation settles down to a chaotic regime, which does not repeat itself and appears irregular and unpredictable. Instead of plotting the time series of the displacement for the Duffing oscillator, we may instead plot the displacement against velocity in the **phase plane** (a 2D **phase space**). Figure 6.3 contains the x–$\dot{x}$ plot of the Duffing oscillator corresponding to the time series of figure 6.2. The dynamical system evolves through time as a continuous curve in the phase plane. The solution **trajectories** for the periodic cases are given in figures 6.3(*a*) and (*b*). The initial conditions are shown as large dots in the figures. The post-transient behaviour of the periodic cases shows up as closed loops in the phase plane, known as **limit cycle attractors**. The winding down of the transient motion to the post-transient limit cycle attractor is particularly obvious for the $x_0 = 3.0$, $\dot{x}_0 = 3.0$ case. The closed loops, corresponding to the post-transient behaviour of the system, may be thought of as a portrait of the periodic attractor. These **phase portraits** are particularly useful in visualizing the dynamics of the system. The phase portrait of the attractor for the chaotic case is given in figure 6.3(*c*). After the initial high frequency 'winding down' of the transient motion, the system settles onto the post-transient attractor for the chaotic solution. This attractor is a much more exotic object than the limit cycles of the periodic cases. Known as a **strange** or **chaotic attractor**, it represents all the possible final states of the chaotic solution. A few cycles of the trajectory around the strange attractor are shown in the phase portrait on the right of figure 6.3(*c*). (Examples of phase portraits obtained from experimentally observed chaotic motion are given in appendix B, figures B.2–B.4.)

We can simplify phase portraits further by sampling the signal under investigation at intervals equal to the characteristic period of the system. By plotting a point in phase space every characteristic or forcing period, rather than the whole phase space trajectory, the phase diagram may be simplified, showing up features that would otherwise be hidden in the complexity of the phase portrait. Thus, if the attractor is periodic with a period equal to that of the forcing period then only one point will be plotted over and over again—once per period of forcing. Figure 6.4(*a*) illustrates this for the limit cycle attractor of figure 6.3(*a*). The large dot in the figure is plotted once per revolution of the trajectory around the limit cycle. Similarly, figure 6.4(*b*) plots sampled points of the period 3 attractor of figure 6.3(*b*). This time we see three sampled points, as the system takes three periods of the forcing function to return to its original dynamical state. Figure 6.4(*c*) plots the sampled points for the chaotic solution: this time the trajectory (shown in the right-hand plot of figure 6.3(*c*)) is omitted for clarity. This plot of the sampled points is known as a **Poincaré map**. Each point is mapped onto the next through the action of the dynamical system in phase space. However, unlike the maps we investigated in chapter 5, this time we are not sure about the exact form of the mapping function required to iterate the points on the map. There is a marked difference in the appearance of the Poincaré map for the chaotic trajectory and the simple repeating dots of the periodic cases. In fact, the points in the Poincaré map for the chaotic case never reappear at the same location, because if they did the system would then repeat. (Examples of Poincaré maps obtained from experimentally observed chaotic motion are given in appendix B, figures B.1 and B.2.)

The chaotic oscillations of the Duffing oscillator are seemingly unpredictable. However, unlike a random number sequence, they are not truly unpredictable. Each time we begin a chaotic oscillation on a specific set of initial conditions we obtain identical time series (provided that exactly the same integration procedure is used). The system is said to be **deterministic** as we can precisely determine the long term behaviour

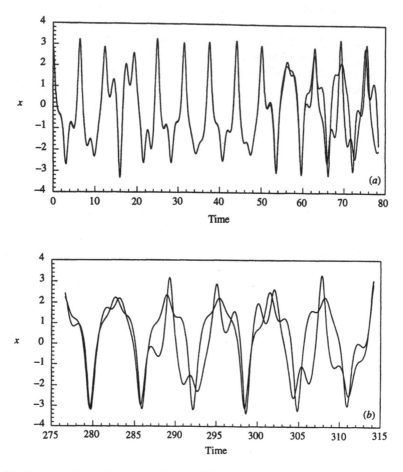

Figure 6.5. Sensitive dependence on initial conditions. (*a*) The drifting apart of two initially very close points on the chaotic time series. Control parameters: $r = 0.05$, $A_f = 7.5$. Initial conditions: $x = 3.00$, $\dot{x} = 0.00$ and $x = 3.01$, $\dot{x} = 0.00$. (*b*) Although similarities occur in the time trace, the long term behaviour is quite different for the two initially close time series.

of a chaotic system if we know the initial conditions exactly. However, and here is the important point, a slight change in the initial conditions of the chaotic system soon results in an entirely different time series. This phenomenon is known as **sensitive dependence on initial conditions** and is a hallmark of chaotic motion. This property is highlighted in figure 6.5 which contains two displacement–time series for the Duffing oscillator with very close initial conditions. One oscillation begins at $x_0 = 3.0$, $\dot{x}_0 = 3.0$ and the other at $x_0 = 3.01$, $\dot{x}_0 = 3.0$. As can be seen from the figure, the two solutions initially appear to follow an identical path (figure 6.5(*a*)). However, as time passes, the solution paths begin to diverge, leading to completely different long term behaviours (figure 6.5(*b*)). Over small scales, the drifting apart of the two solutions is in fact exponential. This divergence property may be quantified using characteristic exponents known as Lyapunov exponents.

These will be discussed in detail in chapter 7.

Sensitivity to initial conditions is what causes the seemingly unpredictable, long-term evolution of chaotic motion, because even a tiny error in the measurement of the initial conditions of a real dynamical system leads rapidly to a lack of predictability of its long-term behaviour. As we cannot measure any real dynamical system with infinite precision the long term prediction of chaotic motion in such systems is impossible, even if we know their equations of motion exactly.

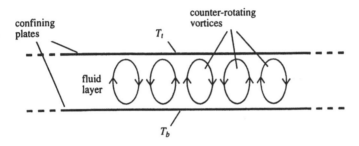

Figure 6.6. A schematic diagram of Rayleigh–Benard convection.

6.3 Chaos in the weather: the Lorenz model

The sensitive dependence on initial conditions of chaotic systems that we observed above in the chaotic response of the Duffing oscillator is more popularly known as the **butterfly effect**. This phenomenon was first discovered by Edward Lorenz during his investigation into a system of coupled ODEs used as a simplified model of 2D thermal convection, known as Rayleigh–Benard convection. These equations are now called the Lorenz equations, or **Lorenz model**. Figure 6.6 shows a schematic diagram of Rayleigh–Benard convection between two horizontal plates. The bottom plate is at a temperature T_b which is greater than that of the top plate, T_t. For small differences between the two temperatures, heat is conducted through the stationary fluid between the plates. However, when $T_b - T_t$ becomes large enough, buoyancy forces within the heated fluid overcome internal fluid viscosity and a pattern of counter-rotating, steady recirculating vortices is set up between the plates. Lorenz noticed that, in his simplified mathematical model of Rayleigh–Benard convection, very small differences in the initial conditions blew up and quickly led to enormous differences in the final behaviour. He reasoned that if this type of behaviour could occur in such a simple dynamical system, then it may also be possible in a much more complex physical system involving convection: the weather system. Thus, a very small perturbation, caused for instance by a butterfly flapping its wings, would lead rapidly to a complete change in future weather patterns. The Lorenz equations are

$$\dot{x} = -\sigma \, (x - y)$$
$$\dot{y} = -xz + rx - y \qquad\qquad (6.3)$$
$$\dot{z} = xy - bz$$

This system has two nonlinearities, the xz term and the xy term, and exhibits both periodic and chaotic motion depending upon the values of the control parameters σ, r and b. σ is the Prandtl number which relates the energy losses within the fluid due to viscosity to those due to thermal conduction; r corresponds to the dimensionless measure of the temperature difference between the plates known as the Rayleigh number; and b is related to the ratio of the vertical height of the fluid layer to the horizontal extent of the convective rolls within it. Note also that the variables x, y and z are not spatial co-ordinates but rather represent the convective overturning, horizontal temperature variation, and vertical temperature variation respectively.

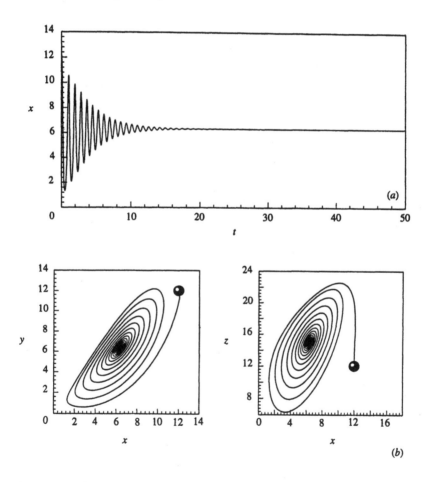

Figure 6.7. A non-oscillatory, post-transient solution to the Lorenz system. (*a*) x–t plot. (*b*) x–y and x–z phase portraits.

Following Lorenz, we set $\sigma = 10.00$ and $b = 2.67$ respectively and make r the adjustable control parameter. Varying the value of r reveals a critical value at $r_c = 24.74$ at which the behaviour of the system changes dramatically. Below r_c the system decays to a steady, non-oscillating, state. Figure 6.7 shows the time series and the x–y and x–z

phase portraits for the system with $r = 16$. Again, the initial condition is shown as a large dot in the phase plane. The rapid decay of the oscillations to a stable fixed point is obvious from both the time series and phase portraits.

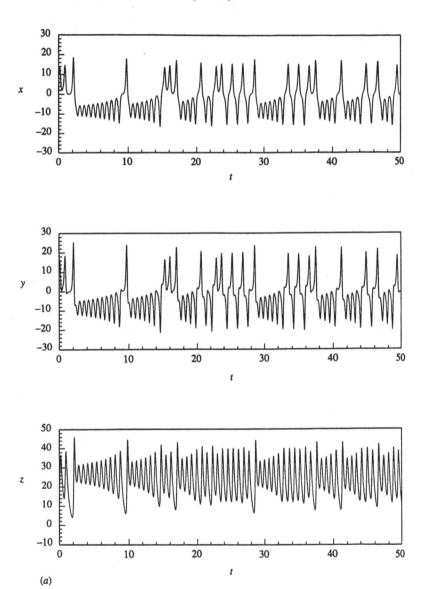

(a)

Figure 6.8. The Lorenz strange attractor. (a) Time series for the Lorenz attractor in chaotic mode. (b) 2D projections of the strange attractor in the phase plane. (c) 3D plot of the attractor in phase space. (Continued overleaf.)

Once r increases beyond r_c, continuous oscillatory behaviour occurs. A value of $r = 28.00$ produces aperiodic behaviour which Lorenz called 'deterministic nonperiodic

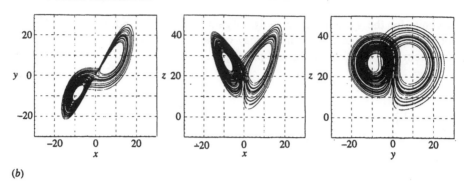

(b)

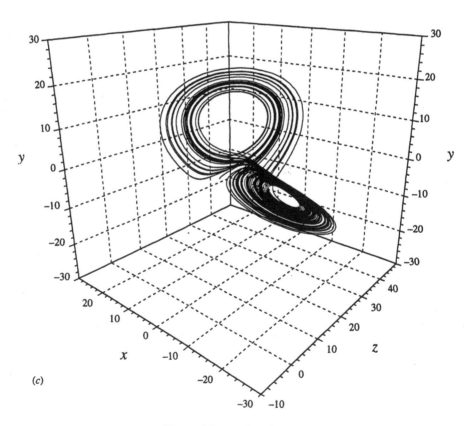

(c)

Figure 6.8. Continued.

flow' and which we now refer to as chaos. Figures 6.8(a)–(c) contain the chaotic system's time series and phase portraits in 2D- and 3D phase space. The post-transient trajectories of the chaotic solution lie on a strange attractor. We see from the figure that the trajectories wind around two distinct lobes of the attractor. After a number of revolutions around one lobe of the attractor the trajectory then switches to the other lobe. It then spirals around this lobe before switching back to the first. The number of revolutions that the trajectory will make around each lobe, before returning to the other, is unpredictable.

The structure of the Lorenz strange attractor is extremely complex. The attractor seems to consist of two surfaces and these appear to merge. However, the solution trajectory can never cross itself, as this would lead to a recurrence of a former state of the dynamical system and would therefore imply a periodic cycle. To achieve non-crossing trajectories, we require a minimum of three dimensions for the phase space, so that the trajectories can avoid themselves. For the same reason, what appear to be merging sheets on the attractor cannot, in fact, be 2D sheets. Upon closer examination these surfaces reveal themselves to be an infinite number of interleaved sheets, having a Cantor-set-like construction. This **fractal structure** is a fundamental property of strange attractors and is another characteristic of chaotic motion. The Lorenz attractor has a fractal dimension of approximately 2.05. In chapter 7, we will look at ways of estimating the fractal dimension of strange attractors.

One way of simplifying the dynamics of the chaotic Lorenz system is to plot the maximum z value, denoted z_n, against the next maximum, denoted z_{n+1}. This is done in figure 6.9(a). We see that the points lie along an approximately 1D map function, which has a gradient greater than unity everywhere upon it and hence is unstable everywhere upon it. Iterating the map should lead to a chaotic sequence of iterates (refer back to chapter 5). The tent map of figure 6.9(b) provides an idealization of the Lorenz map. Graphical iteration of the tent map is shown schematically. The Lorenz map has reduced the 3D dynamical system to a 1D (i.e. it requires only one variable—z) non-invertible map. That the map is only an approximation to the true dynamics becomes apparent when we consider that the Lorenz equations are invertible and may be run forward or backwards in time to produce unique solutions, but, as we learned in chapter 5, a non-invertible map cannot be run backwards in time to produce single unique solutions. In fact, the Lorenz map does have a finite thickness, and may only be used to gain an approximation to the dynamics of the system.

Increasing the control parameter r still further above r_c we find series of reverse bifurcations from the chaotic state to periodic orbits. One such periodic solution is shown in figure 6.10(a)–(c) for $r = 100.5$. The Lorenz equations posses an important symmetry. If $[x(t), y(t), z(t)]$ is a solution to the Lorenz equations, then so is $[-x(t), -y(t), z(t)]$. Therefore, Lorenz attractors are either symmetric or have a symmetric partner. Figure 6.10(d) contains the symmetric partner to the solution given in figure 6.10(a)–(c). The symmetry property is obvious when comparing figure 6.10(b) with 6.10(d).

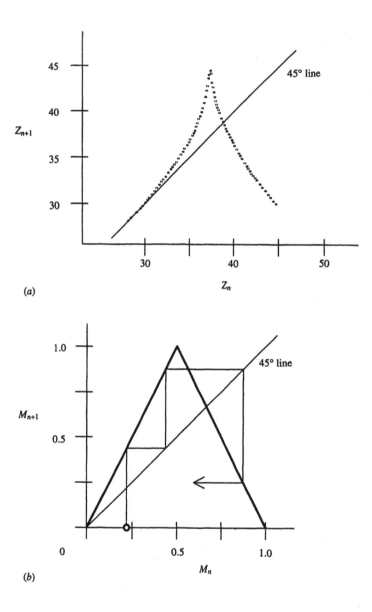

(a)

(b)

Figure 6.9. The Lorenz map. (a) Maxima of the Z variable plotted against subsequent maxima. (b) The idealized tent map function.

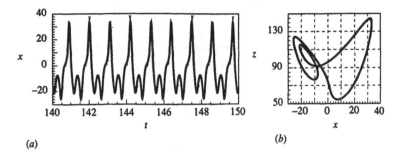

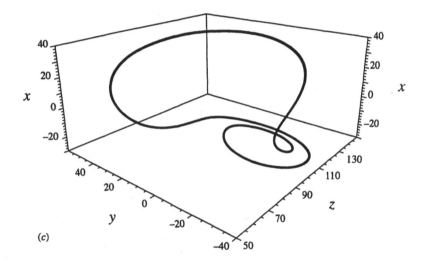

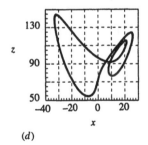

Figure 6.10. A periodic attractor for the Lorenz system with $\sigma = 10.00$, $b = 2.67$, and $r = 100.5$. (a) The post-transient x–time series. (b) The x–z phase portrait. (c) The phase space trajectory. (d) The symmetric partner of the solution of (a)–(c).

6.4   The Rössler systems

The Rössler system was proposed as a dynamical system simpler than the Lorenz model, but which could still exhibit chaos. Its equations of motion are

$$\dot{x} = -(y - z)$$
$$\dot{y} = x + ay \tag{6.4}$$
$$\dot{z} = b + xz - cz$$

The Rössler system has only one nonlinear term, the xz term, whereas the Lorenz model has two nonlinear terms. Setting the system parameters $a = 0.2$ and $b = 0.2$, and varying the third parameter, c, we can find a period doubling sequence to chaos. In figure 6.11(a), four phase portraits are shown for the Rössler system for four different values of c. The $c = 2.0$ case has a single limit cycle attractor; when c is increased to 3.5 we see that the trajectory loops twice before reconnecting and repeating itself; for $c = 4$ the trajectory loops four times around before reconnecting and repeating itself. This period doubling sequence continues to chaos, e.g. for $c = 5.7$. The time series for the chaotic attractor is shown in figure 6.11(b) from which we see that the trajectories spend most of their time very close to the x–y plane, i.e. with the z co-ordinate very close to zero, and the z component 'fires' intermittently.

As with periodic attractors, strange attractors attract the solution trajectories of the dynamical system to a certain set of possible final states, this means that they exist in particular bounded regions of phase space. However, we have also seen that chaotic motion has the property of sensitive dependence on initial conditions where nearby trajectories separate exponentially. The question is then, 'How do strange attractors manage to attract on large scales and at the same time repel on smaller scales?' (i.e. 'How do they remain bounded on large scales while trajectories diverge on small scales?'). They accomplish these seemingly contradictory properties through a process of **stretching and folding** of the solution trajectories on the attractor. We saw the divergence of nearby trajectories for the Duffing oscillator in figure 6.5, and this corresponds to localized stretching on the attractor at small scales. Similarly, nearby trajectories on the Rössler attractor diverge exponentially. At larger scales the attractor exhibits folding and the Rössler attractor is particularly good at illustrating the folding process (see figure 6.11(c)). If we pick a trajectory near to the edge of the attractor at position A we can follow its movement around the edge of the base of the attractor until it is lifted up, when the z component 'fires'. The trajectory is then folded over and reinjected into the attractor (location B), ending up near to the centre of the attractor. After another revolution on the attractor the trajectory is now somewhere within the main band of the attractor at position C. In general, nearby trajectories separate at an exponentially fast rate, 'stretching', and then they are folded up and reinjected back into the attractor. This causes the mixing of the trajectories in the attractor and is the mechanism which leads to the rapid decorrelation of initially nearby points on the attractor making accurate long term prediction impossible (although at least we know that the future state will still be somewhere on the attractor). The Rössler strange attractor is often referred to as the folded band due to the obvious nature of the folding part of the mixing process.

We saw previously that we need at least a 3D phase space for a chaotic attractor. Now we turn our attention to a system which requires a 4D phase space. This dynamical system, also produced by Rössler, is described by the following four equations:

$$\dot{x} = -y - z$$
$$\dot{y} = x + ay + w$$
$$\dot{z} = b + xz$$
$$\dot{w} = cz + dw$$

(6.5)

where chaos may be observed for $a = 0.25$, $b = 3.0$, $c = -0.5$ and $d = 0.05$. The form of chaos in this system is known as **hyperchaos**, as the stretching may take place in two principal directions on the attractor. Figure 6.12 shows the attractor for the chaotic hyperchaos equations in xyw phase space, with the time evolution of the z variable plotted below. The z variable acts as a threshold variable to limit the outwards growth

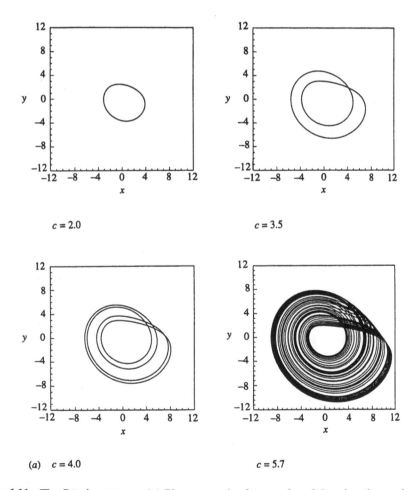

$c = 2.0$

$c = 3.5$

(a) $c = 4.0$

$c = 5.7$

Figure 6.11. The Rössler system. (a) Phase portraits for $a = b = 0.2$ and various values of c. (b) Time series for the chaotic case with $a = b = 0.2$ and $c = 5.7$. (c) Phase space trajectory of the Rössler strange attractor for $a = b = 0.2$ and $c = 5.7$. (Continued overleaf.)

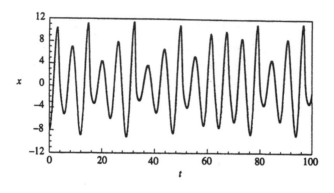

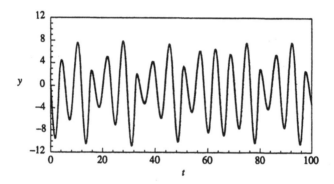

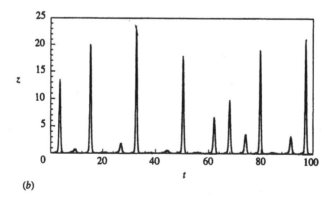

(b)

Figure 6.11. Continued.

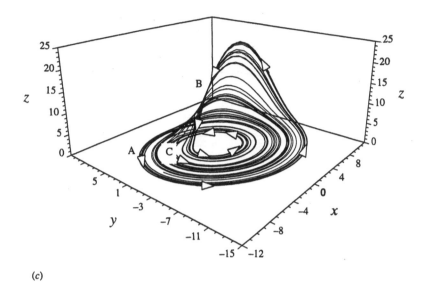

(c)

Figure 6.11. Continued.

of the trajectory in phase space. The intermittent firing of the z variable acts to reinject the trajectories back into the main region of the strange attractor. Rössler put forward his hyperchaos model as the next step in the search for more complex models which would allow for richer behaviour in their dynamics.

6.5 Phase space, dimension and attractor form

In the present context of dynamical systems, it is easy for the meaning of dimension to become confused. This is, therefore, a good place to put into context the various meanings of dimension that may arise. Let us begin by defining **degrees of freedom**. The number of degrees of freedom in a dynamical system is defined to be the number of independent variables required to describe the instantaneous state of the system (hence, phase space is often referred to as state space). Therefore, the number of degrees of freedom required for a full description of a dynamical system is simply the number of initial conditions that are required to specify the system. Phase space is defined as a *'mathematical space with orthogonal co-ordinate directions representing each of the variables needed to specify the instantaneous state of the system'*. The number of degrees of freedom is then the minimum dimension required for the phase space of the system. For the Lorenz and Rössler models we require a knowledge of three initial conditions and therefore a 3D phase space is required to fully describe the dynamics of these systems. Similarly, the Rössler hyperchaos model requires a 4D phase space. What about the Duffing oscillator? It has a different form from the coupled set of first-order ODEs of the Lorenz and Rössler models and, in addition, the time variable appears explicitly in the Duffing oscillator, i.e. it is a non-autonomous system. First it may be useful to rewrite

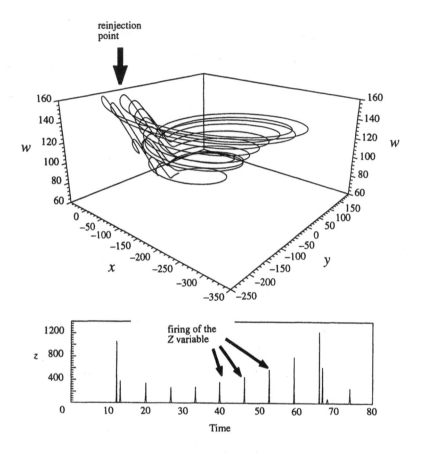

Figure 6.12. The strange attractor for the Rössler hyperchaos attractor.

the Duffing oscillator in the following form:

$$\dot{x} = y$$
$$\dot{y} = A_f \cos z - r\dot{x} - x^3 \qquad (6.6)$$
$$\dot{z} = 1$$

where the introduction of the new variables y and z has decomposed the system into a set of first-order ODEs. Furthermore, the auxiliary variable z has removed the explicit time dependence of the system; in other words, the system is now autonomous, just like the Lorenz and Rössler equations. Since the three variables x, y and z fully describe the system, we now see that a 3D phase space is required for the system.

The spatial dimension in which the dynamical system exists is generally not equal to the phase space dimension. The Duffing oscillator models a dynamical system which is constrained to move in only one spatial dimension, the x direction. The Lorenz system models convection in 2D space, where the variables x, y and z are related to properties

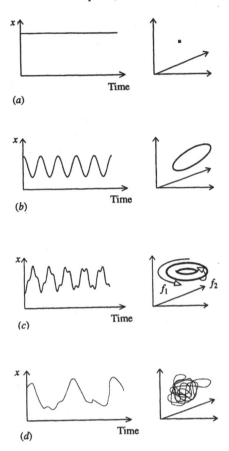

Figure 6.13. A schematic diagram of attractor types with corresponding time series. (*a*) A steady state solution with a fixed point attractor. (*b*) A single period solution with a limit cycle attractor. (*c*) Two incommensurate frequencies with a torus attractor. (*d*) A chaotic solution with a strange attractor.

of the system. The Rössler system does not model a real system at all, but is rather a simplified model based on the Lorenz system. We see, therefore, a distinct difference between the dimension of phase space and the spatial dimension of the problem, although both are integer dimensions.

The dimension of the attractor in phase space is important and generally distinct from the spatial dimension of the system and the dimension of the phase space. The dimension of the attractor is related to its geometric form. For instance, a non-oscillatory solution corresponds to a fixed point in phase space which attracts nearby trajectories. We have already seen this type of behaviour for the Lorenz system in figures 6.7(*a*) and (*b*), where the trajectories spiralled into the fixed point solutions, figure 6.13(*a*) shows this schematically. The dimension of a steady state point attractor is zero. Once oscillations begin to appear in the system, the attractor may take on different forms. A single period solution has a limit cycle attractor, shown schematically in figure 6.13(*b*). Again we have already come across this type of attractor (figures 6.4(a) and 6.11(a)). The closed curve of the limit cycle has a dimension of one. Once a second frequency emerges in the solution of a dynamical system, then the two frequencies wind round one another on a torus, or doughnut shape, in phase space. This is shown schematically in figure 6.13(c), where the oscillation of frequency f_2 winds around the f_1 oscillation in phase space. If the two frequencies are incommensurate (i.e. the second frequency is not rationally related to the first) then the trajectory completely fills up the surface of a torus in phase space. The surface of the torus has a dimension of two. If more frequencies appear in the system, then higher-dimensional tori are created requiring higher-dimensional phase spaces. Finally, for chaotic motion, strange attractors appear which have non-integer, fractal dimensions (figure 6.13(d)).

As an example of the different dimensions involved in specifying the dynamics of nonlinear systems consider the Lorenz system. It models convection in <u>two</u> spatial dimensions, has <u>three</u> degrees of freedom, and hence requires a <u>three</u>-dimensional phase space; and may, under certain parameter settings, develop a chaotic attractor (figure 6.8(c)) with a <u>fractal</u> dimension slightly greater than two.

6.6   Spatially extended systems: coupled oscillators

Coupled oscillator systems are often used to model complex dynamical systems in which interactions take place between spatially distinct oscillators. For example, chains of coupled oscillators have been used to model the behaviour of platoons of road vehicles on the highway. The behaviour of these extended systems depends on their spatial structure and the method of coupling employed: i.e. we may link adjacent oscillators together in a chain (1D), grid (2D) or lattice (3D); link each oscillator to every other; or perhaps even form rings of oscillators. In addition, the coupling itself may be unidirectional, bi-directional or multi-directional. There are many possibilities and here we will look briefly at selected output of two oscillator chains. A schematic diagram of an oscillator chain is given in figure 6.14(a). The behaviour of the chain under the action of external forcing, F, depends on the form of coupling employed. In figures 6.14(b) and (c), the equations of motion for the oscillator chain for two different forms of coupling are presented, together with chaotic output for the parameter values given. Both systems are based on the Duffing oscillator (6.1). Notice, however, that they have the damping parameter r and the forcing frequency ω as adjustable control parameters (compare to equation (6.2)). The chain in figure 6.14(b) has unidirectional coupling: that is the nth oscillator in the chain is driven by the output of the $(n-1)$th oscillator. Moving along the chain, the time series becomes more complex and shows quite clearly the emergence of higher frequency components. The corresponding phase portraits are given and the Duffing phase portrait is recognizable for the $n = 1$ oscillator. The oscillators in the second system, given in figure 6.13(c), are coupled in both directions along the chain, i.e. the nth oscillator is influenced by both the $(n-1)$th oscillator and the $(n+1)$th oscillator. We can see from the time series that this coupling does not lead to higher-frequency components dominating the behaviour of the system. In addition, this bi-directional coupling causes the output of the nth oscillator to be dependent upon the number of oscillators in the chain, N. (This is not the case for unidirectional coupling.) This is highlighted in figure

coupling

forcing
F

(a)

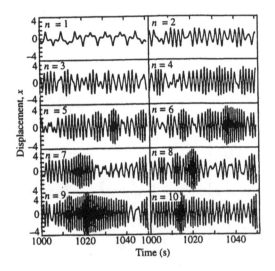

Time series for ten coupled oscillators

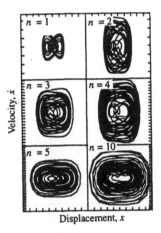

Phase portraits from the time
series for oscillator numbers
1–5 and 10

Unidirectional coupling equations of motion:

$$\ddot{x}_n + r\dot{x}_n + x_n^3 = x_{n-1} \quad n = 1,2,3,\ldots,N \text{ and } x_0 = \cos(\omega t); \ r = 0.044\,96; \ \omega = 0.4496$$

(b)

Figure 6.14. Coupled oscillator systems. *(a)* A schematic diagram of a coupled oscillator chain. *(b)* Unidirectional coupling. *(c)* Bi-directional coupling. (After Addison (1995). Reproduced by permission of the publisher, Academic Press Limited, London. © Academic Press Limited.) (Continued overleaf).

6.14*(c)* by the two separate sets of time series and phase portraits for an $N = 6$ chain and an $N = 10$ chain. Notice that the output for the first six oscillators is quite different for the same control parameter settings and initial conditions. As mentioned above, there are many other ways of coupling oscillator systems and the final structure and complexity of the equations of motion depends upon the characteristics of the system being modelled. By linking together oscillators the number of degrees of freedom of the dynamical system increases. Thus, the dimension of phase space required increases and correspondingly these systems may exhibit hyperchaos with high dimensional strange attractors.

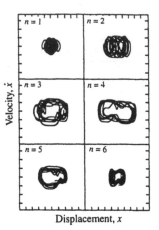

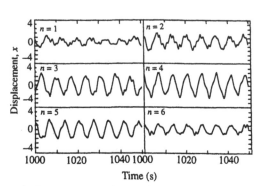

Time series for six coupled oscillators

Phase portraits from the time series for oscillator numbers 1–6

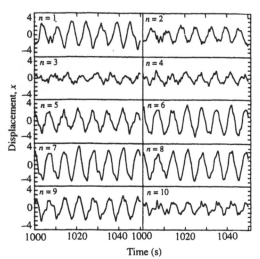

Time series for ten coupled oscillators

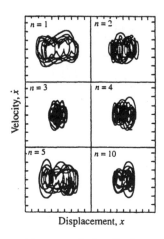

Phase portraits from the time series for oscillator numbers 1–5 and 10

Bidirectional coupling equations of motion:

$$\ddot{x}_1 + r\dot{x}_1 + (x_1 - x_2)^3 = -\cos(\omega t)$$

$$\vdots \qquad \vdots \qquad \vdots \qquad \vdots$$

$$\ddot{x}_n + r\dot{x}_n + (x_n - x_{n+1})^3 = (x_{n-1} - x_n)^3 \qquad n = 1,2,3,...,N \text{ and } r = 0.044\,96; \ \omega = 0.4496$$

$$\vdots \qquad \vdots \qquad \vdots \qquad \vdots$$

$$\ddot{x}_N + r\dot{x}_N + (-x_N)^3 = (x_{N-1} - x_N)^3$$

(c)

Figure 6.14. Continued

6.7 Spatially extended systems: fluids

Many real systems can be modelled as sets of ODEs which are then solved numerically to produce solutions; just as we did earlier in the chapter for the Duffing oscillator, Lorenz system, Rössler systems and the coupled oscillators. However, it is not always possible to describe real dynamical systems so simply. Often physical systems are <u>continuously</u> spatially distributed and require partial differential equations (PDEs) of motion. These continuous systems are infinite dimensional, as they require an infinite number of initial conditions distributed through space to define them. (Notice that the coupled oscillator systems described in the last section were also spatially distributed, however, those systems contained discrete oscillators coupled together, i.e. they were <u>discretely</u> spatially distributed, and hence were finite dimensional requiring only a series of coupled ODEs to describe their motion.) The dynamical behaviour of a continuously spatially extended system may vary both in space and time, leading to observations of regular and chaotic spatio-temporal oscillations. In this section we concentrate on one class of spatially extended systems, fluid systems, where chaos has shed light upon the transition from the ordered (laminar) state to a highly disordered (turbulent) state.

A fluid is a non-rigid, continuous interconnected mass, which in general may exhibit either laminar or turbulent flow. Laminar flows are characteristic of slow-moving or highly viscous flows where the fluid particles move in an ordered fashion, sliding over themselves in sheets (or laminae, hence 'laminar'). As an example of a laminar flow, think of the slow-moving viscous flow from a spilled honey pot on a table with a slight incline. Turbulent flows, on the other hand, are characteristic of fast-moving or low-viscosity flows, where small disturbances in the flow blow up causing the fluid particles to move in an unpredictable fashion, mixing themselves up from one point in the flow to the next. As an example of a turbulent flow, think of the highly energetic flow in the plunge pool at the base of a waterfall. Fluid flow is governed by a set of nonlinear PDEs, the Navier–Stokes equations, written in full as

$$\rho \left[\frac{\partial u}{\partial t} + u \frac{\partial u}{\partial x} + v \frac{\partial u}{\partial y} + w \frac{\partial u}{\partial z} \right] = -\frac{\partial p}{\partial x} + \mu \left[\frac{\partial^2 u}{\partial x^2} + \frac{\partial^2 u}{\partial y^2} + \frac{\partial^2 u}{\partial z^2} \right] + F_x$$

$$\rho \left[\frac{\partial v}{\partial t} + u \frac{\partial v}{\partial x} + v \frac{\partial v}{\partial y} + w \frac{\partial v}{\partial z} \right] = -\frac{\partial p}{\partial y} + \mu \left[\frac{\partial^2 v}{\partial x^2} + \frac{\partial^2 v}{\partial y^2} + \frac{\partial^2 v}{\partial z^2} \right] + F_y \quad (6.7a)$$

$$\rho \left[\frac{\partial w}{\partial t} + u \frac{\partial w}{\partial x} + v \frac{\partial w}{\partial y} + w \frac{\partial w}{\partial z} \right] = -\frac{\partial p}{\partial z} + \mu \left[\frac{\partial^2 w}{\partial x^2} + \frac{\partial^2 w}{\partial y^2} + \frac{\partial^2 w}{\partial z^2} \right] + F_z$$

and the continuity equation (here we adopt the simpler case of an incompressible fluid)

$$\left[\frac{\partial u}{\partial x} + \frac{\partial v}{\partial y} + \frac{\partial w}{\partial z} \right] = 0. \quad (6.7b)$$

In these equations, p is the fluid pressure; u, v and w are the velocity components in the x, y and z directions; ρ is the density and μ is the dynamic viscosity of the fluid; and F is the external body force on the fluid. These PDEs show the dependence of the fluid system on both space and time. The equations are nonlinear, with the nonlinearity residing in the convective acceleration terms $u \, \partial u / \partial x$, $v \, \partial u / \partial y$ etc. In addition, they model a continuous, or infinite dimensional dynamical system, quite unlike the finite dimensional systems we have encountered so far.

As was stated above, flowing fluids exist in either of two states: a laminar state, where the fluid flows in an orderly and predictable fashion; or a turbulent state, where the fluid particles move in a disorderly fashion and rapid decorrelation is evident within the flow field in both space and time. Much attention in the literature has focused on the possibility of explaining the transition from the laminar state to turbulence in terms of the transition from regular to chaotic motion. We might reasonably expect that the simpler dynamical systems encountered in the preceding sections will have nothing to tell us about the much more complex problem of fluid turbulence. However, recent experiments have found that, for certain strictly controlled flow problems, infinite dimensional fluid systems may exhibit regular and chaotic motion of low dimension. In other words, they may behave as systems with only a few degrees of freedom.

The key dynamical control parameter for a fluid system is the Reynolds number. This is the ratio of inertial to viscous forces in the flow and is defined as

$$Re = \frac{\rho V D}{\mu} \tag{6.8}$$

where ρ is the density of the fluid; V is a characteristic flow velocity; D is a characteristic length scale of the problem; and μ is the dynamic viscosity of the fluid. Below a critical value of the Reynolds number, Re_{crit}, the viscous forces dominate and the flow is laminar. Above Re_{crit}, inertial forces dominate and the flow is turbulent. As the Reynolds number is increased, a transition from laminar to turbulent flow takes place. This transition may appear either suddenly or gradually depending upon the geometry of, and the presence of noise in, the system. With careful variation of the Reynolds number, chaos may be observed at the transition between laminar and turbulent flow in some fluid systems.

It is possible to derive a mathematical approximation to the Navier–Stokes equations in the form of a truncated set of ODEs, in much the same way as the Lorenz equations were formed. One such set of ODEs by Boldrighini and Francheschini describing the motion of a 2D incompressible fluid on a torus is

$$\dot{x}_1 = -2x_1 + 4x_2x_3 + 4x_4x_5$$
$$\dot{x}_2 = -9x_2 + 3x_1x_3$$
$$\dot{x}_3 = -5x_3 - 7x_1x_2 + Re \tag{6.9}$$
$$\dot{x}_4 = -5x_4 - x_1x_5$$
$$\dot{x}_5 = -x_5 - 3x_1x_4$$

where Re is the Reynolds number of the system. It is hoped that this type of approximation may possess some of the essential features of the full Navier–Stokes equations for low Reynolds numbers, although at present it is not known for certain whether the approximation does possess these features. Figure 6.15 shows two phase portraits of the strange attractor for these truncated equations with Re = 33.00. Each phase portrait is a 2D snapshot of the attractor. Just as a photograph is a 2D image of a 3D world, here we have 2D snapshots of a 5D world. There are ten different phase portraits that could be plotted for the attractor in the phase plane, not counting mirror images.

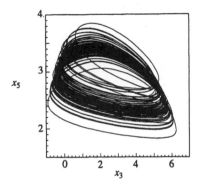
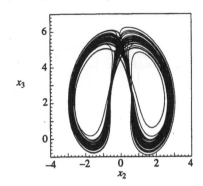

Figure 6.15. Two phase portraits of the strange attractor of the truncated Navier–Stokes equations for Re = 33.00.

6.7.1 Taylor–Couette flow

The transition from the laminar to the turbulent state in a real fluid may be rapid or gradual (controllable) depending very much upon the characteristics of the flow problem. Chaos theory has provided new insights into the transition between laminar and turbulent fluid flows for certain experimental fluid systems. The most extensively studied of fluid systems has possibly been Taylor–Couette flow, which occurs in the fluid between two concentric cylinders rotating at rates independent of each other. Figure 6.16(a) contains a schematic diagram of the Taylor–Couette experimental apparatus. Again, the Reynolds number is the control parameter of the system, although variations in the physical dimensions of the apparatus also affect the behaviour of the system. At the inner and outer cylinder walls the fluid velocity is exactly equal to the wall velocity, this is known as the no-slip boundary condition, and for low Reynolds numbers the velocity of the fluid changes through a uniform gradient between the cylinder walls. Above a critical value of the Reynolds number, the flow organizes itself into discrete, counter-rotating vortices in the gap between the walls. The number of tubular, doughnut shaped vortices (known as Taylor vortices) that appear in the gap depends very much on the geometry of the gap itself. A cross section of the gap, filled with vortices, is drawn in figure 6.16(a). At higher Reynolds numbers, a travelling wave appears in the vortices, and this travels around the annulus at a speed related to that of the inner cylinder (figure 6.16(b)). A further increase in the Reynolds number leads to the appearance of a second travelling wave. Increasing the Reynolds number still higher leads to a breakdown in the flow to a chaotic state. The fractal dimension of the strange attractor for the chaotic state increases approximately linearly with increasing Reynolds number beyond the onset of chaos at Re_{chaos}: this is shown schematically in figure 6.16(c). This low dimensional chaotic state of a dynamical system is often referred to as **weak turbulence**; as distinct from the **fully developed turbulence** typical of fluid systems at high Reynolds numbers, which consists of high dimensional spatio-temporal behaviour and which is still not fully understood by the scientific community. (Appendix B contains experimental time series from a Taylor–Couette experiment which exhibit both periodic and chaotic behaviour.)

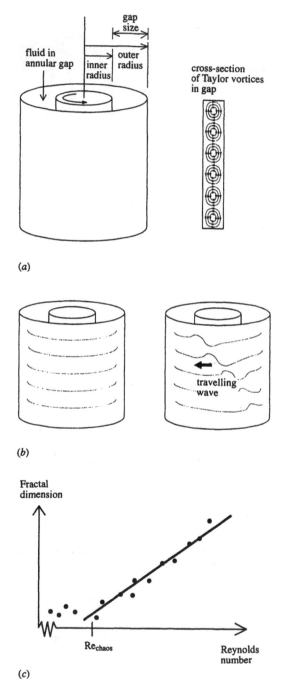

Figure 6.16. The Taylor–Couette experiment. (See also appendix B for experimental output.) (*a*) The apparatus and cross section of vortices in the gap. (*b*) Schematic diagram of the development of a travelling wave. (*c*) The behaviour of the dimension against Reynolds number in the chaotic region.

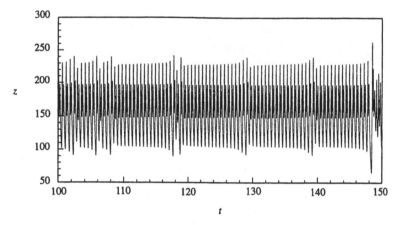

Figure 6.17. Intermittency in the Lorenz system for $r = 166.3$.

6.8 Mathematical routes to chaos and turbulence

Investigations of nonlinear dynamical systems, both theoretical and experimental, have shown that both relatively simple low-dimensional systems and highly complex infinite dimensional systems may evolve from a steady state to a chaotic state as a control parameter is increased. Much theoretical interest has been centred around the routes taken from the steady to chaotic states of such systems, and whether these routes may be realized both in simple models and in infinitely complex spatio-temporal systems. There is now known to be a large number of possible routes to chaos in dynamical systems. We concentrate here on four mathematical descriptions of the route taken by fluid systems from the steady state, through the transition to the chaotic state and on to a fully developed turbulent state, whatever that might be. All routes begin with a transition from an initial non-oscillatory, stationary state (refer back to figure 6.13(*a*)) to a singly periodic orbit (figure 6.13(*b*)) and this transition is known as a Hopf bifurcation. The **routes to chaos and turbulence** are as follows.

(i) *The Landau model.* Landau was the first to suggest a mathematical route to turbulence. He proposed that from an initial laminar state an oscillation would appear in the flow at a critical value of Reynolds number. Increasing the Reynolds number would soon lead to another oscillation entering the system, with a frequency incommensurate with the first. Further increases in Re would lead to further oscillations of incommensurate frequencies appearing in the system. The final, fully developed turbulence would occur only at infinite Reynolds number and would consist of a quasi-periodic motion of great complexity.

(ii) *The Ruelle–Takens route.* Ruelle and Takens have conjectured that a strange attractor will appear at the third bifurcation of the Landau sequence. That is from the laminar state an initial oscillation will set in at the critical Reynolds number, then, at a higher Reynolds number, another incommensurate oscillation will appear, causing a torus to form in phase space (see figure 6.13(*c*)). However, rather than a third distinct oscillation appearing, a chaotic attractor is more likely to appear in the flow. This is the route described above for the Taylor–Couette flow.

(iii) *The Feigenbaum period doubling route.* We have already come across the Feigenbaum route to chaos in the logistic map of chapter 5, and the Rössler system (figure 6.11(a)), whereby successive period doublings result in a strange attractor at a finite value of the control parameter.

(iv) *The Manneville–Pomeau intermittent route.* In this case the system alternates between a singly periodic solution and chaos. An illustration of **intermittency** occurring in the Lorenz system is given in figure 6.17 for the control parameter set at $r = 166.3$. The system exhibits long periods of almost periodic behaviour interspersed with shorter periods of chaotic behaviour. The ratio of chaotic to periodic behaviour in the time series increases with r. Intermittency is also found in the logistic map.

The Landau model is now not regarded as a realistic route to turbulence as it has not appeared in any experimental situation. In addition the following should be noted.

(i) Routes (ii)–(iv) have all been found in real systems. In particular, all three have been observed in fluid convection experiments. (Rayleigh–Benard convection. See also appendix B.)

(ii) The Feigenbaum number has been found in many hydrodynamic and electronic systems which exhibit chaos.

(iii) The intermittent route may be further subdivided into three distinct types (not described here).

(iv) Deviations from the above routes have been found in some experiments. For example, in free surface wave experiments, the Faraday experiment, departures from the period doubling route (i.e. period $= 1, 2, 4, 8, 16, 32, \ldots$) have been found: these are $1, 2, 4, 12, 14, 16, 18, 20, 22, 24, 28, 35, \ldots$ and $1, 2, 3, 4, 6, 12, 16, 18, 24, 28, 30, 32, 36, \ldots$.

(v) Hysteresis has been observed in some experimental routes, that is the route taken by the system as the control parameter is increased is different from the route taken as the control parameter is decreased.

(vi) Other routes have now been found experimentally.

6.9 Chapter summary and further reading

6.9.1 Chapter keywords and key phrases

Duffing oscillator	*transient/post-transient*	*basin of attraction*
attractor	*oscillation*	*trajectories*
phase plane	*control parameter*	*strange/chaotic attractor*
limit cycles	*phase space*	*hyperchaos*
Poincaré maps	*phase portrait*	*fully developed turbulence*

sensitive dependence on   *deterministic dynamics*   *routes to chaos*
initial conditions   *fractal structure*   *intermittency*
(butterfly effect)   *(of a strange attractor)*   *degrees of freedom*
Lorenz model   *stretching and folding*
Rössler system   *weak turbulence*

6.9.2 General

In this chapter we have taken a brief look at the occurrence of chaotic behaviour in continuous dynamical systems. The brevity of the treatment is compensated for in part by the extensive list of references given below which should allow the curious reader to deepen his or her knowledge in a specific field of interest.

The text by Scheck (1990) covers basic dynamics, leading on to an account of chaotic motion. General introductions to chaos are given in the texts by Berge *et al* (1984), Schuster (1988), Marek and Schreiber (1991), Tsonis (1992), Ott (1993), Çambel (1993), McCauley (1993), Strogatz (1994), Korsch and Jodl (1994) and Kaplan and Glass (1995). (The Korsch and Jodl text comes with extensive software for simulating many chaotic systems.) The four volumes by Abraham and Shaw (1982, 1983, 1984, 1988) give an excellent visual treatment of the geometry of the dynamics of nonlinear systems: see also the comprehensive paper by Bradley (1995). Deeper mathematical insights into nonlinear oscillations and chaos can be found in the books by Hagedorn (1981), Guckenheimer and Holmes (1983), Mira (1987), Jordan and Smith (1987), Sagdeev *et al* (1988), Seydel (1988), Wiggins (1988, 1990), Devaney (1989), Ruelle (1989) (a concise introduction), Lakshmanan and Daniel (1990), Lichtenberg and Lieberman (1992), Ingraham (1992), Palis and Takens (1993) and Glendinning (1994). In this chapter we have concentrated exclusively on dissipative dynamical systems. However, there is a category of non-dissipative systems, known as conservative or Hamiltonian systems, which are useful in describing certain dynamical systems where there is no dissipation, or it is so slight that it can be ignored, e.g. models of the solar system. The occurrence of chaos in Hamiltonian systems is detailed by Ozorio de Almeida (1988), Zaslavsky *et al* (1991) and Seimenis (1994).

6.9.3 The Duffing oscillator

The Duffing oscillator has received much attention from the dynamics community. Ueda (1979, 1980a) used the form of the Duffing equation given by equation (6.2) to examine chaos in an electrical circuit with a nonlinear inductor. A full description of the effect of the control parameters r and A_f on the long term behaviour of the Duffing oscillator is given by Ueda (1980b). Many other variants of the Duffing oscillator have been used to model a wide variety of physical systems: the most prominent of these is one with both a linear and a cubic restoring force term, i.e. $\ddot{x} + r\dot{x} - s_1 x + s_2 x^3 = A_f \cos \omega t$, which models the behaviour of a **buckled cantilever beam** or an **electron in a plasma** (Moon and Holmes 1979 (see appendix B), Moon 1980, Brunsden *et al* 1989, Asfar and Masoud 1992). The text by Moon (1987) contains details for the construction of a simple buckled beam experiment which can be modelled using the Duffing equation. Gottwald *et al* (1992) have produced a 'cart on a track' experimental system which can mimic the dynamics of the Duffing oscillator. Thompson and Stewart (1986) provide many more details on the Duffing oscillator, together with a particularly lucid description of the dynamics of the Rössler system. Moon and Li (1985a) present the fractal structure of the basins of attractions for periodic responses to the Duffing oscillator. In a separate paper, the same authors present dimension measurements from experimental and numerical Duffing models (Moon and Li 1985b). Other variants of the Duffing equation include modified forcing (Wiggins 1987, Kapitaniak 1988a, Addison *et al* 1992, Bapat 1995), the addition of noise (Kapitaniak 1986, Falsone and Elishakoff 1994), modified damping (Ravindra and Mallik 1994, Xie and Hu 1995) and coupled Duffing oscillators (Burton and Anderson 1989, Leung and Chui 1995, Addison 1995 (see figure 6.14)). Further information on the properties of the Duffing oscillator may be found in the following papers by Szemplinska-Stupnicka (1987, 1994), Vaneck (1994), Pezeshki and Dowell (1987), Dowell and Pezeshki (1986, 1988), Rahman and Burton (1986), Ishii *et al* (1986), Holmes (1979) and Tseng and Dugundji (1971).

6.9.4 The Lorenz system

There are two references which stand out amongst a whole host of others for the Lorenz system: these are Lorenz's original 1963 paper and the comprehensive treatment of the Lorenz equations by Sparrow (1982). (Arguably, Lorenz's remarkable paper stands out amongst the whole literature on chaos.) Lorenz (1963) details the derivation of his model from the convective equations of Saltzman (1962). The value of $r = 28.00$ used in this chapter has no special qualities other than that it was the value used by Lorenz to produce chaos. Many other values of r produce chaotic motion: see for instance the book by Sparrow (1982), who provides many details of the workings of the Lorenz system, including reverse bifurcation sequences which occur with increasing r. There is now a very large number of papers which investigate properties of the Lorenz system and its variants, e.g. those by White and Tongue (1995), Chern (1994), Alvarez-Ramirez (1994), Sarathy and Sachdev (1994), Chen *et al* (1994) and Liu and Barbosa (1995). A literary analogy to the butterfly effect is given by Nave (1994). Chaos has been observed experimentally in the turbulent convection of fluid between two plates with a temperature difference—the Rayleigh–Benard experiment (Giglio *et al* 1981, Arneodo *et al* 1983): see also the collection of papers on chaos in **thermal convection** edited by Bau *et al* (1992). In addition, all three of the valid routes to chaos have been observed in this system (Gollub and Benson 1980) (see appendix B).

6.9.5 The Rössler systems

Rössler (1976) gives stereoscopic images of the Rössler strange attractor and Lorenz strange attractor. The article by Rössler (1979) contains the hyperchaos equations, again with a stereoscopic image of the strange attractor. One interesting application of hyperchaotic systems has been in the area of secure communications, where the increased complexity of hyperchaotic systems has a beneficial effect in masking transmitted messages. The technique is illustrated by Peng *et al* (1996) with both the Rössler hyperchaos system and a discrete hyperchaos system formed from coupled maps, and Cuomo *et al* (1993) with **Lorenz-based circuits** (see also Cuomo and Oppenheim (1993), Kocarev and Parlitz (1995) and Mirasso *et al* (1996)).

6.9.6 Fluids and other spatially extended systems

Background texts in fluid mechanics and turbulence which also include some chaos theory include those by Tritton (1988), Chevray and Mathieu (1993) and Frisch (1995) (who presents the logistic map as a poor man's Navier–Stokes equations). The collection of articles by Kadanoff (1993) contains some good introductory essay papers on the onset of chaos in hydrodynamic (and much simpler) systems. Collections of articles and papers concerning turbulence and chaotic phenomena in fluids are edited by Swinney and Gollub (1981), Barenblatt *et al* (1983) and Tatsumi (1984).

Information on the similarities and dissimilarities of **turbulent intermittency** in pipe flows and intermittency in simpler dynamical systems described by sets of ODEs can be found in the papers by Sreenivasan and Ramshankar (1986) and Huang and Huang (1989). **Chaotic vortex shedding** has been found in the wake of a cylinder by Elgar *et al* (1989) and in the wake of an airfoil by Williams-Stuber and Gharib (1990). Chaos has also been observed in a **closed flow vortex system** by Tabeling *et al* (1990). Analytical treatments of **vortex motions** are to be found in the papers by Aref *et al* (1989), Conlisk *et al* (1989), Novikov (1991) and Noack and Eckelmann (1992). More information on chaos in **Taylor–Couette flow** is given by Gollub and Swinney (1975), Brandstäter *et al* (1983), Brindley (1986), Brandstäter and Swinney (1987) and Mullin and Price (1989). The key references of the **routes to chaos and turbulence** are those by Landau (1944), Ruelle and Takens (1971), Feigenbaum (1978, 1980) and Pomeau and Manneville (1980) (see also Keolian *et al* (1981) who describe a departure from the period doubling route.) Hopf bifurcations begin all four routes to turbulence, however, this is but one among a whole range of bifurcation types that nonlinear systems may exhibit. For examples of other bifurcation types and alternative routes to chaos see many of the general chaos texts cited above; or more specifically consult those by Hale and Kocak (1991) or Kuznetsov (1995). Libchaber (1984) lists ten references which give details on experiments involving the onset of chaotic behaviour.

Although chaos has been found in low Reynolds number flows, at present it is not known whether high Reynolds number, fully developed turbulence is some form of spatio-temporal chaos with a large, but not infinite, number of degrees of freedom. Some thoughts on the similarities with chaos in low-dimensional and high-dimensional turbulent fluid systems are given by Ruelle (1980, 1983a, b), Guckenheimer (1986), Miles (1983) and Lanford (1981), see also the article by Deissler (1986) for chaos in the Navier–Stokes equations. The fractal nature of high Reynolds number turbulent flows is discussed by Mandelbrot (1984a) and Constantin *et al* (1991), and its multifractal nature by Benzi *et al* (1984), Meneveau and Sreenivasan (1987, 1991), Sreenivasan and Prasad (1989), Frisch and Orszag (1990), Frisch and Vergassola (1991) and Borgas (1993a, b). Low-dimensional chaos in the coherent structures within turbulent boundary layers is detailed by Aubry *et al* (1988); and coherent structures arising out of **turbulence in the atmosphere** are reviewed in a collection of papers edited by Nezlin (1994). An unsuccessful attempt to find low-dimensional chaos in highly turbulent channel flows is detailed by Brandstäter *et al* (1986).

Another area of fluid mechanics which has received much attention from the chaos community recently is on the application of chaos theory to the **mixing of fluids** through chaotic advection. A detailed discussion of this topic is given in the collection of papers by Aref and El Naschie (1995); see also the book by Ottino (1989). More information, including the route to chaos, on the **truncated Navier–Stokes equations** is given by Boldrighini and Franceschini (1979) and Franceschini and Tebaldi (1979); see also Foias and Treve (1981) for the truncation required to give long term solutions to the Navier–Stokes equations. Information on other forms of spatially extended systems, e.g. chemical, physiological, and electrical, exhibiting both spatial and temporal oscillations, is given in the collection of papers edited by S C Müller *et al* (1994) and that edited by Artuso *et al* (1991).

A relatively simple way to produce an infinite dimensional system is to incorporate a time delay. The most famous example is the Mackey–Glass equation (Mackey and Glass 1977) which models **blood cell production**. Delay equations are infinite dimensional as they require each delay variable to be completely defined over the delay time interval. Numerical solutions to delay equations inevitably involve their reduction to approximated, discretized and hence finite-dimensional systems, although it is assumed that such simulations may accurately represent the behaviour of the original infinite dimensional system (Farmer 1982).

There is also much interest centred around the dynamics of discrete spatially extended systems, in particular, the dynamics of discretely located systems that are coupled together. This has been done for systems of coupled maps: see for instance the collection of articles devoted to these coupled map lattices edited by Kaneko (1992); see also Kaneko (1989, 1990) and Beck (1994). Matthews *et al* (1991) go one step further and examine coupled continuous time oscillators, citing many references on their application in biology and physics. Bifurcation and chaos in selected coupled oscillators is detailed by Awrejcewicz (1991). Rosenblum *et al* (1996) detail the coupling of Rössler systems, and Kocarev and Parlitz (1996) describe a Lorenz system coupled to a Rössler system, see also the articles by Yamada and Fujisaka (1984) and Heagy *et al* (1994) and the references cited elsewhere in this section, e.g. coupled Duffing oscillators, following vehicles on a roadway, etc. Another class of discrete spatially extended dynamical systems attracting much attention are cellular automata (Wolfram 1984, Pires *et al* 1990, Manneville *et al* 1989). These mathematical models essentially consist of a grid upon which an initial distribution of dynamical particles (automatons) are left to interact with each other according to specific rules. Bak and Chen (1989) and Bak *et al* (1990) present a cellular automata, forest fire model as a spatially extended system which results in fractal dissipation and compare it to the phenomenon of turbulence. Wolfram (1986) derives hydrodynamic equations from a class of cellular automata.

6.9.7   Miscellaneous subject areas

There are many collections of articles and papers on the theory and applications of chaos in the natural and applied sciences: see for instance those edited by Helleman (1980), Haken (1981), Kalia and Vashishta (1982), Garrido (1983), Campbell and Rose (1983), Iooss *et al* (1983), Kuramoto (1984), Chandra (1984), Cvitanovic (1984), Fischer and Smith (1985), Sarkar (1986), Holden (1986), Bai-Lin (1984, 1987, 1989), Berry *et al* (1987), Lundqvist *et al* (1988), Kim and Stringer (1992), Kapitaniak (1992), Tong and Smith (1992), Mullin (1993) and Harrison *et al* (1996). These texts contain papers covering practically all aspects of numerate study. More specific subject areas and topics are dealt with below.

Those interested in chaos in **biology and medicine** are referred to the introductory texts by Glass and Mackey (1988), West (1990) and Nicolis (1991), the collections of papers by Degn *et al* (1987) and Hoppensteadt (1979), and the excellent review article by Olsen and Degn (1985). Bélair *et al* (1995) are the guest editors for a collection of papers on the **dynamics of human illness** and Denton *et al* (1990), Kyriazis (1991) and Renshaw (1994) describe chaotic dynamics in the context of **cardiology, gerontology** and **biometry** respectively. Schmid and Dünki (1996) look at nonlinearity in the **human EEG** and Femat *et al* (1996) find chaos in a **human cardiac signal**. **Ecologists** are referred to the text by Pahl-Wostl (1995). **Geologists** and **geophysicists** are introduced to the applications of both fractal geometry and chaotic dynamics by Turcotte (1992). More specifically the chaotic dynamical behaviour of **earthquakes** is investigated by Carlson and Langer (1989) and Huang and Turcotte (1990b), where individual earthquake faults are modelled by chains of coupled oscillators with a static–dynamic friction law. Various aspects of the theory and applications of chaos in **quantum systems** are brought together in a collection of articles in the texts edited by Casati (1985) and Gay (1992): see also Steeb and Louw (1986), Gutzwiller (1990) and Nakamura (1993), and the texts specifically concerning **quantum optics** by Arecchi and Harrison (1987) and Abraham *et al* (1988). Ogorzalek (1995) and Chua and Hasler (1993) are guest editors for collections of papers devoted to nonlinear dynamics of **electronic systems**; see also the article by Kilias *et al* (1995) for some applications of **electronic chaos generators**. A substantial collection of papers edited by Madan (1993) is devoted to chaos in a simple electric circuit, known as Chua's circuit (This circuit, together with two distinctive forms of strange attractor, is detailed in appendix B). This relatively simple circuit enables simultaneous numerical, mathematical and experimental investigation of its dynamics. Chaos in coupled circuits is described by Rulkov and Sushchik (1996). **Structural engineers** are pointed towards the text by El Naschie (1990), and a structural analysis of a magnetoelastic ribbon is provided by Moorthy *et al* (1996). Further papers of interest in this area are those concerned with chaos in **elastic–plastic beams** (Poddar *et al* 1988), **thin shells** (Dekhtyaryuk *et al* 1994), **aeroelastic fluttering** of thin panels (Reynolds *et al* 1993) and **flow induced vibrations** of cylinders (Plaschko *et al* 1993). Those with an interest in **naval architecture** should see the study of nonlinear behaviour of **articulated offshore platforms** by Choi and Lou (1991), or that of **capsize phenomena** by Soliman and Thompson (1991). **Mechanical engineers** interested in chaos arising out of **rotating machinery** should see for example the articles by Neilson and Gonsalves (1992), Zhao and Hahn (1993), Brown *et al* (1994) and Adiletta *et al* (1996a, b): the latter two papers contain experimentally observed chaos in a **journal bearing** and a **jeffcot rotor** respectively. (Figure B.2 in appendix B contains various plots for the journal bearing system.) **Traffic engineers** interested in the possibilities of chaos arising out of **following vehicles** on a roadway should see for instance the articles by Jarrett and Xiaoyan (1992), P C Müller *et al* (1994) or Addison and Low (1996); see also that by Liu *et al* (1996), who discuss chaos in **road vehicle driver steering control**. **Chemists and chemical engineers** are referred in the first instance to the introductory text of Scott (1994), and at a more advanced level to the book by Scott (1991): see also the review articles by Doherty and Ottino (1988) and Villermaux (1993). Chaos has been observed in the **Belousof–Zhabotinsky (BZ) chemical reaction** (Hudson *et al* 1981, Turner *et al* 1981, Roux and Swinney 1981, Roux 1983, Roux *et al* 1983, Dolnik and Epstein 1996). Appendix B contains periodic and chaotic output for the BZ chemical reaction in the form of time series, frequency spectra and attractor plots. There are many texts on the role of chaos in **economic dynamics**: simple introductions to the topic are given in the texts by Parker and Stacey (1994) and Goodwin (1990); more advanced treatments are given by Leydesdorff and van den Besselaar (1994) and Lorenz (1993). In addition to these texts, there are many collections of articles on the subject; see for example the collections of papers edited by Barnett *et al* (1989), Creedy and Martin (1994), and Pesaran and Potter (1993). Also, the research paper of Hommes (1989) may be of interest. An interesting text by Shaw (1984) details an experiment which investigates the route to chaos of the dripping sequence from a **faucet** (tap) as the flow rate is increased; see also the article by Néda *et al* (1996). Chaotic bubbling is described by Tritton and Egdell (1993), and an elementary text by Baker and Gollub (1990) concentrates on **pendulum dynamics** as it explains many of the concepts of chaotic dynamics. **Astronomers** are pointed towards the collection of papers edited by Roy and Steves (1995), and the articles by Sussman and Wisdom (1988, 1992), which present numerical evidence that the motions of both **Pluto** and the **solar system** as a whole are chaotic.

Further references citing chaos in miscellaneous subject areas are, in no particular order: Paladino (1994) on chaos in **wastewater treatment** reactors, Fisher (1993) on applications of chaos theory to **haematology**, Testa *et al* (1982) on chaos in **driven nonlinear semiconductors**, Allen· *et al* (1991) on chaos in **geostrophic flows**, Ott and Tél (1993) on **chaotic scattering**, Ikeda *et al* (1980) on **optical chaos** (more details in appendix B), Paidoussis and Moon (1988) and Paidoussis *et al* (1989) on **fluid elastic vibrations** of a flexible pipe conveying fluid, Waldner *et al* (1985) on irregular routes to chaos in **pumped ferromagnets**, Middleton (1990) on the potential applications of chaos in the **earth sciences**, Lauterborn and Parlitz (1988) on applications of chaos physics in **acoustics** (see also Lauterborn (1996)), Barnett and Chen (1988) on the application of chaos theory to **economic inferences**, Olsen and Degn (1977) on chaos in an **enzyme reaction**, Gibbs *et al* (1981) on **optical chaos**, Labos (1987) on chaos in **neural networks**, Tritton (1986) on chaos in a **spherical pendulum**, Grabec (1986) on chaotic oscillations arising in **mechanical cutting processes**, Hassell *et al* (1991) on chaos in **insect population dynamics**, Ge *et al* (1996) on chaos in a **model gyroscope**, Yagisawa *et al* (1994) on chaos in **gel–liquid-crystal phase transitions**, Mercader *et al* (1990) on **hysteresis** in a model chaotic system, Velarde and Antoranz (1981) and Colet and Braiman (1996) on chaos in a **model laser system**, Barndorff-Nielsen *et al* (1993) on **networks and chaos**, Nettel (1995) on **waves, solitons and chaos**, Harrison and Biswas (1986) and Gioggia and Abraham (1983) on chaos in **real lasers**, Collins and Stewart (1993) on chaos in coupled oscillator models of **animal gaits**, Chatterjee and Mallik (1996) on chaos in **oscillators with impact damping**, Mayer-Kress *et al* (1993) on the **bassoon-like musical sounds** generated from a chaotic circuit and Dabby (1996) on the generation of **musical variations** using the Lorenz equations.

6.9.8 Numerical methods

The numerical solutions to the nonlinear systems in this chapter were computed using a fourth-order **Runge–Kutta algorithm** (Johnson and Riess 1982, Press *et al* 1992, Stuart and Humphries 1996). The FORTRAN program GENERAL given in appendix A at the end of this book produces a numerical solution to the Lorenz equations using the fourth-order Runge–Kutta method. However, you may find it easier to use one of the many sophisticated mathematical software packages around, most of which will allow you to produce numerical solutions to systems of ODEs. Addison *et al* (1992) briefly describes the Runge–Kutta fourth-order method in a discussion of the accuracy of the solutions obtained by numerical methods to nonlinear equations, in particular a modified Duffing oscillator.

6.10 Revision questions and further tasks

Q6.1 List the keywords and key phrases of this chapter and briefly jot down your understanding of them.

Q6.2 Using Newton's second law, derive the equation of motion for the Duffing oscillator.

Q6.3 Explain, in your own words, the phenomenon of the butterfly effect in chaotic dynamical systems. Include in your answer an explanation of why the butterfly effect leads to an inability to predict the long-term evolution of <u>real</u> dynamical systems when they behave chaotically.

Q6.4 An experimenter finds that as he/she increases the control parameter of the fluid dynamical system under investigation, the following frequencies are observed: initially no frequency, i.e. steady, laminar flow, then a frequency of 17.20 Hz appears followed by 8.60 Hz, then 4.30 Hz and then 2.15 Hz.

(*a*) What route do you think the system is taking to turbulence?

(*b*) What are the three main routes via chaos to turbulence that have now been observed in real dynamical systems?

Q6.5 Use the output from the program GENERAL in appendix A to generate time series and phase portraits of the strange attractor of the Lorenz system for parameter values of $\sigma = 10$, $b = 2.67$ and $r = 28$. (You may also use other mathematical software available to you to do this.) Plot x, y and z against time. Then plot the x–y, x–z and y–z phase portraits.

Q6.6 Keeping σ and b constant at 10 and 2.67 respectively, investigate the dynamics of the Lorenz system for r over the range 15 to 300.

Q6.7 Use the output from the program GENERAL in appendix A to generate time series and phase portraits of the Rössler system. Do this for $a = b = 0.2$, and vary c from 2 to 5.7

Q6.8 Modify the program GENERAL to produce numerical solutions to the Rössler hyperchaos system and the truncated Navier–Stokes equations given in the text.

Q6.9 Modify the program GENERAL to produce numerical solutions to the Duffing oscillator given by equation (6.2).

Q6.10 Further, modify the program GENERAL to produce numerical solutions to the coupled oscillator systems detailed in section 6.6 and figure 6.14. Try various chain lengths and parameter values. (Additionally, you could try your own form of coupling.)

Chapter 7

Characterizing chaos

7.1 Introduction

Chaotic motion has many defining properties and these may be examined using various techniques. The problem lies with finding methods which allow us to differentiate between purely random (i.e. noisy or unpredictable) time series and those which are chaotic (i.e. seemingly unpredictable). In this chapter we investigate the characteristics of chaotic motion which enable us to both quantify it and set it apart from periodic and random motions.

7.2 Preliminary characterization: visual inspection

Before using any sophisticated form of numerical technique to analyse a time series, one should first **visually inspect** the series to see if any particularly noticeable features are present. For instance, from a visual inspection of a velocity–time trace of a flowing fluid one can distinguish between laminar and turbulent flow. In chapter 6, we saw time series for both periodic and chaotic solutions to many dynamical systems. For example, figures 6.2(a) and (b) contained periodic solutions to the Duffing oscillator and figure 6.2(c) contained a chaotic solution. Looking back at the figures one can tell, simply by looking at them, that the periodic solutions do indeed appear to repeat their behaviour along the time series, whereas the chaotic solution does not appear to repeat itself (at least not within the segment of time series plotted in the figure). Visual inspection of the phase portraits (figure 6.3) and Poincaré sections (figure 6.4) of the Duffing oscillator once again shows obvious periodic behaviour.

Visual inspection is a useful initial tool and it is always good practice to visually inspect time series data before employing a more complex analysis technique. (It is also a particularly effective way of spotting outrageously silly mistakes in computed time series.) However, visual inspection does have some major drawbacks, which are summarized as follows.

(i) It is not always possible to inspect a large enough segment of time series to determine whether or not the series actually repeats itself.
(ii) It is difficult to visually determine whether a signal composed of many commensurate frequencies is indeed periodic.

(iii) If the time series is composed of as few as two incommensurate frequencies, then it may be impossible to determine visually whether the time series is periodic or chaotic.

(iv) A dominant frequency in a signal may mask other frequencies or a slightly chaotic motion.

7.3   Preliminary characterization: frequency spectra

The Fourier transform enables the frequencies present in a time series to be inspected. The popularity of using Fourier transforms in signal analysis has increased due to the development of a fast Fourier transform (FFT) algorithm for computers, which has enabled discrete Fourier transforms to be calculated with exceptional speed. In this section, we will only briefly skim over the salient points of Fourier analysis: for a more comprehensive account the reader is directed to the references given at the end of this chapter. As we will deal solely with discretized time series in this chapter we concentrate on the discrete Fourier transform of the discrete time series, x_j, $j = 0, 1, 2, 3, \ldots, (N-1)$, defined as

$$X_k = \frac{1}{N} \sum_{j=0}^{N-1} x_j e^{-i2\pi kj/N} \qquad k = 0, 1, 2, \ldots, (N-1) \qquad (7.1a)$$

where i is the imaginary number $\sqrt{-1}$. (Note that the definitions of i, j, k and N will change throughout this chapter—beware!) The time spacing between each sampled point must be such that its reciprocal (i.e. the sampling frequency) is at least twice the maximum frequency to be resolved in the signal. The inverse transform of (7.1a) is

$$x_j = \sum_{k=0}^{N-1} X_k e^{i2\pi kj/N} \qquad j = 0, 1, 2, \ldots, (N-1). \qquad (7.1b)$$

By plotting the Fourier transform of a time series, periodic, multiply periodic, quasi-periodic and non-periodic signals may be differentiated.

Figure 7.1 contains the frequency spectra associated with the time series of periodic and chaotic oscillations of the Duffing oscillator (refer back to figures 6.2–6.4). The attractors are plotted next to the frequency spectra. The frequency spectrum for the periodic case (figure 7.1(a)) is composed of two peaks, one at the forcing frequency, f_f ($= 1/2\pi$) and one at one third of the forcing frequency. This is consistent with the time series (figure 6.2(b)) and Poincaré section (figure 6.4(a)) in which we saw a period 3 oscillation occurring in the system. The chaotic spectrum of figure 7.1(b) has distinct peaks at the forcing frequency and its harmonics, $3f_f$, $5f_f$ and $7f_f$. In addition to this, a **broad-band spectrum** of Fourier components have now appeared amongst the harmonics. In deterministic systems, this spectral broadening is a hallmark of the onset of chaos. Often, instead of a frequency spectrum, a power spectrum is plotted. The power spectrum is the squared magnitude of the frequency spectrum. Figure 7.1(c) contains the power spectrum of the chaotic case in figure 7.1(b). The vertical scale is logarithmic, and we can see that the logarithmic power spectrum plot shows more clearly the broad-band spectral components interspersed with higher frequency components.

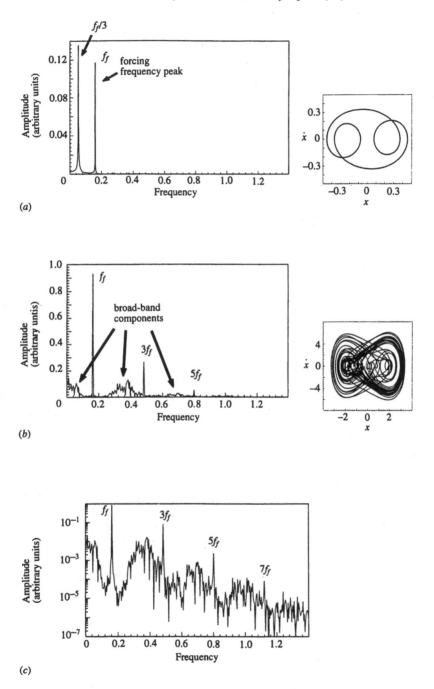

Figure 7.1. The Fourier spectra for the Duffing oscillator and the Lorenz system. (a) The frequency spectrum of the periodic attractor containing two peaks. (b) The frequency spectrum of the chaotic attractor containing dominant harmonic peaks and broad-band components. (c) The power spectrum of (b). (d) The power spectra for the x and z time co-ordinates in the Lorenz model in chaotic mode. (Continued overleaf.)

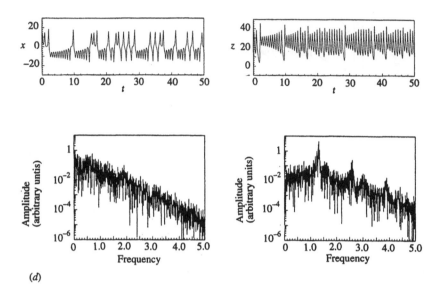

(d)

Figure 7.1. Continued.

Two power spectra for the Lorenz model in chaotic mode are given in figure 7.1(d), (refer back to figure 6.8). One for the x–time series and one for the z–time series (shown above the spectra). Both plots contain broad-band spectral components, however, there is a dominant peak in the z spectrum at a frequency of around 1.35, corresponding to the orbital frequency of the trajectory around the two lobes of the attractor. This peak is not evident in the other spectrum. The z–time series is slightly simpler in form as it does not contain the inflection points, present in the x–time series, corresponding to the trajectory crossing from one attractor lobe to the other. The two spectra illustrate the ability to gain more information by choosing a 'better' dynamical variable to study. Visual inspection of the time series helps here, on the other hand, if time permits, it is always a good idea to investigate the frequency spectra (and other characteristics) of all the variables in the system. Power spectra from various experimental dynamical systems exhibiting both periodic and chaotic motion are given in appendix B, figures B.3 to B.7.

Broad-band spectral components may also arise from random signals (or noise). However, we know that the Duffing oscillator is a deterministic system and includes no random component in its equations of motion. We may, therefore, deduce from a broad-band frequency spectrum that the system is exhibiting chaotic motion. If, on the other hand, we have a time series from a real dynamical system which may contain noise, then it is more difficult to differentiate between the broad band components associated with noise, and those associated with chaotic behaviour. Thus, the frequency spectra are generally not enough to confirm the presence of chaos in an experimental signal, which will inevitably contain an element of background noise. Therefore, we require techniques which can differentiate between those time series which are random (or noisy) and those which are chaotic. We can do this either by measuring the exponential divergence of nearby trajectories on the attractor or by investigating its fractal structure. In the next two sections, methods are described which perform these measurements.

7.4 Characterizing chaos: Lyapunov exponents

As we saw in chapter 6, nearby trajectories diverge exponentially on strange attractors, giving rise to the butterfly effect in chaotic dynamical systems. We can measure this exponential divergence and characterize it using **Lyapunov exponents**. A schematic diagram of two separating trajectories is given in figure 7.2(*a*). In this figure, ε_0 is the

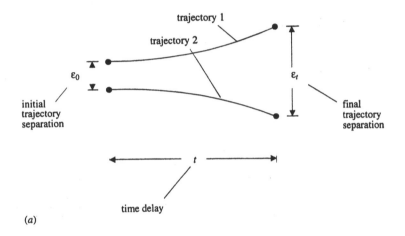

(*a*)

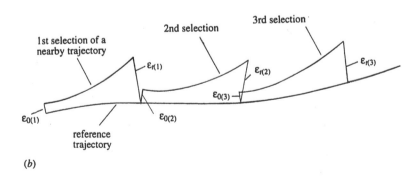

(*b*)

Figure 7.2. The Lyapunov exponent. (*a*) Definition sketch of trajectory separation for the Lyapunov exponent. (*b*) Finding the average Lyapunov exponent over the attractor through the tracking of nearby trajectories. (*c*) Divergence and convergence. (*d*) Deformation of an infinitesimally small hypersphere in phase space. (Continued overleaf.)

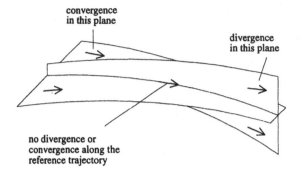

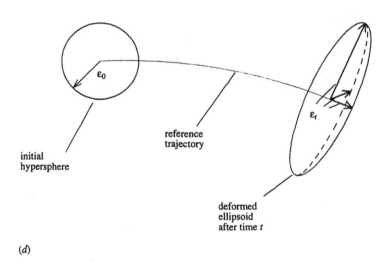

Figure 7.2. Continued.

observed separation of two very close points on separate trajectories on the attractor. After an elapsed time, t, the trajectories have diverged and their separation is now ε_t. As this separation is exponential, we may write

$$\varepsilon_t = \varepsilon_0 \, e^{\lambda t} \qquad\qquad (7.2a)$$

where λ is the Lyapunov exponent. Rearranging, we obtain

$$\lambda = \frac{1}{t} \ln \left[\frac{\varepsilon_t}{\varepsilon_0} \right]. \tag{7.2b}$$

Note that here we have used natural logarithms to define the Lyapunov exponent, however, any base may be used. In practice, base ten and base two are often used, the latter base giving the Lyapunov exponent in terms of bits of information loss per second. The above expression for the Lyapunov exponent is used to measure the average divergence properties of experimental strange attractors. This is done by following a reference trajectory through time and comparing the divergence of 'close by' trajectories over a portion of the attractor cycle time. The method is illustrated in figure 7.2(*b*), where the reference, or fiducial, trajectory is followed and its divergence from a nearby trajectory is monitored. Once the separation becomes too large, another nearby trajectory is selected and so on. The Lyapunov exponent is calculated many times at various locations on the attractor, and an average taken. If we monitor N nearby trajectory segments, the averaged Lyapunov exponent for the attractor is then

$$\bar{\lambda} = \frac{1}{N} \sum_{i=1}^{N} \frac{1}{t} \ln \left[\frac{\varepsilon_{t(i)}}{\varepsilon_{0(i)}} \right] \tag{7.3}$$

although in practice the bar above λ, denoting an averaged quantity, is usually omitted.

As we saw in chapter 6, nearby trajectories on strange attractors diverge, while still remaining in a bounded region of phase space due to the folding process. In addition, trajectories near to the attractor (but not on it) are attracted to it, i.e. they converge on to it (see figure 7.2(*c*)). We have divergence in some directions and convergence in others. To fully characterize the divergence and convergence properties of an attractor we require a set of Lyapunov exponents, one for each orthogonal direction of divergence/convergence in phase space. The number of Lyapunov exponents required to define the attractor is equal to the dimension of its phase space. Chaotic attractors have at least one finite positive Lyapunov exponent. On the other hand, random (noisy) attractors have an infinite positive Lyapunov exponent, as no correlation exists between one point on the trajectory and the next (no matter how close they are), i.e. the divergence is instantaneous. Stable periodic attractors have only zero and negative values of λ. Thus, the Lyapunov exponent is a valuable measure which may be used to categorize chaotic attractors.

If we calculate the Lyapunov exponent for orthogonal directions of maximum divergence in phase space, we obtain a set of Lyapunov exponents $(\lambda_1, \lambda_2, \lambda_3, \ldots, \lambda_n)$, where n is the dimension of the phase space. This set of Lyapunov exponents is known as the **Lyapunov spectrum** and is usually ordered from the largest positive Lyapunov exponent, λ_1, down to the largest negative exponent, λ_n, i.e. maximum divergence to maximum convergence. The Lyapunov spectrum may be found by monitoring the deformation of an infinitesimally small hypersphere of radius ε on the attractor (see figure 7.2(*d*)). Through time, the sphere is stretched in the directions of divergence (positive λs) and squeezed in the directions of convergence (negative λ) taking up the shape of an ellipsoid. The exponential rates of divergence or convergence of the principal axes of this ellipsoid give the spectrum of Lyapunov exponents. One of the exponents in the spectrum is always zero and corresponds to the direction in phase space aligned with the trajectory, which neither expands nor contracts through time.

Often it is enough to denote the Lyapunov spectrum symbolically. Thus (+, +, 0, −) would denote a dynamical system with two positive, one zero and one negative exponent. This system would in fact be hyperchaotic as it is stretching in two principal orthogonal directions. Hyperchaos is defined as a chaotic system which has more than one positive Lyapunov exponent, i.e. it can stretch in two or more orthogonal directions. For high-dimensional systems of equations, Lyapunov spectra have been found with as many as twenty positive exponents. However, only one positive Lyapunov exponent is required for chaos and hence it is sufficient to check whether λ_1 is positive. If it is, then the system is chaotic.

We may also find Lyapunov exponents for non-chaotic attractors. A point attractor has only negative Lyapunov exponents, as all motions spiral into the fixed point, i.e. for a 3D phase space we obtain symbolically (−, −, −). The other possibilities for a 3D phase space are the following: a limit cycle defined symbolically as (0, −, −) where all trajectories spiral onto the limit cycle, and the zero exponent corresponds to the direction along the limit cycle trajectory; an attracting torus defined as (0, 0, −); and a strange attractor defined as (+, 0, −).

The sum of Lyapunov exponents is the average contraction rate of volumes in phase space, i.e. the rate of volume contraction of the ellipsoid in figure 7.2(d). The sum is less than zero in dissipative dynamical systems, as the post transient solutions lie on attractors with zero phase volume, i.e.

$$\sum_{i=1}^{n} \lambda_i < 0 \qquad (7.4)$$

where n is the number of Lyapunov exponents in the spectrum. Hence, we cannot have attractors of the form (+, +, 0) as only positive Lyapunov exponents would lead to exponential growth of the attractor volume. The calculated Lyapunov spectra for the Lorenz strange attractor (using the control parameters for the chaotic case given in chapter 6) have been found to be (2.16, 0.00, −32.4); similarly, for the Rössler strange attractor they are (0.13, 0.00, −14.1); and for the Rössler hyperchaos attractor they are (0.16, 0.03, 0.00, −39.00).

The positive Lyapunov exponents are directly related to the directions of trajectory divergence in phase space. A knowledge of these positive exponents enables us to quantify the limitations of predictions of the future state of the dynamical system based on a knowledge of the current state of the system to a finite resolution. The error inherent within a finite precision measurement blows up exponentially and soon becomes of the order of the attractor itself. As an example, let us consider the simple case of a chaotic attractor in 3D phase space (+, 0, −), e.g. the Lorenz or Rössler system. If we denote the attractor size by ε_a, and the measurement error by ε_e, then the time required for the error to blow up to the size of the attractor is

$$t_p = \frac{1}{\lambda_1} \ln\left[\frac{\varepsilon_a}{\varepsilon_e}\right] \qquad (7.5)$$

where t_p is the prediction horizon, i.e. the time above which predictions are useless.

Notice that t_p increases logarithmically with the precision, ε_e, illustrating why only short term prediction is possible for chaotic systems.

7.5   Characterizing chaos: dimension estimates

7.5.1   Box counting dimension

The dimension of an attractor is a measure of its geometric scaling properties (its 'fractalness') and has been called 'the most basic property of an attractor'. However, as we saw in chapters 2 and 3, many definitions of dimension exist. The simplest, and most easily understood definition of dimension is the **box counting dimension**, D_B, which we came across in equations (3.1)–(3.4). For an attractor of unit hypervolume

$$D_B = \lim_{\delta \to 0} \left[\frac{\log(N)}{\log(1/\delta)} \right] \tag{7.6}$$

where N is the number of hypercubes of side length δ used to cover the attractor. (Note that as with the Lyapunov exponent definition we are free to choose any base we wish for the logarithm. In subsequent examples within this chapter we shall follow usual practice and use base ten for dimension expressions unless otherwise stated.) The dimension of the probing hypercubes is chosen to be equal to the dimension of phase space. Hence, to find the box counting dimension of the Lorenz system, 3D hypercubes (i.e. boxes) are used; whereas for the Rössler hyperchaos system, 4D hypercubes are used. In addition, the above expression assumes an attractor of unit hypervolume, otherwise a more general expression (based on equation (3.4a)) is required, which defines D_B in terms of the slope of the $\log(N)$–$\log(1/\delta)$ plot.

7.5.2   The information dimension

Another frequently cited dimension estimate is the **information dimension**, D_I. As with the box counting dimension, the attractor is covered with hypercubes of side length δ. This time, however, instead of simply counting each cube which contains part of the attractor, we want to know how much of the attractor is contained within each cube. This measure seeks to account for differences in the distribution density of points covering the attractor, and is defined as

$$D_I = \lim_{\delta \to 0} \left[\frac{I(\delta)}{\log(1/\delta)} \right] \tag{7.7a}$$

where $I(\delta)$ is given by Shannon's entropy formula,

$$I(\delta) = - \sum_{i=1}^{N} P_i \log(P_i) \tag{7.7b}$$

where P_i is the probability of part of the attractor occurring within the ith hypercube of side length δ. For the special case of an attractor with an even distribution of points, an identical probability, $P_i = 1/N$, is associated with every box. Hence, $I(\delta) = \log(N)$, and equation (7.7a) reduces to the box counting dimension of equation (7.6). Thus, D_B simply counts all hypercubes containing parts of the attractor, whereas D_I asks how much of the attractor is within each hypercube and correspondingly weights its count.

7.5.3   Correlation dimension

The calculation of D_B and D_I, in practice, often requires a prohibitive amount of computation time. Therefore, the most widely used dimension estimate for attractors is the **correlation dimension**, D_C, which is computationally efficient and relatively fast when implemented as an algorithm for dimension estimation. To define D_C we first need to define the correlation sum, C_r, as follows:

$$C_r = \frac{1}{N(N-1)} \sum_{i=1}^{N} \sum_{j=1; j \neq i}^{N} \theta\left(r - |X_i - X_j|\right) \tag{7.8a}$$

where θ is the Heaviside function, r is the radius of an n-dimensional hypersphere centred on each sampled point on the attractor trajectory, X_i, $i = 1, 2, 3 \ldots N$. Here we are using the compact notation X_i to denote the multi-dimensional vector that is the ith phase space co-ordinate of the attractor, i.e. for an attractor in 3D phase space with co-ordinates x, y and z, $X_i = (x_i, y_i, z_i)$.

It is worth spending a little time on a physical interpretation of equation (7.8a). A definition sketch is given in figure 7.3(a), where the hypersphere is centred on one of the points defining the attractor trajectory, X_i. The number of other points on the attractor within the hypersphere are counted. This is, in fact, what the Heaviside function does in the above equation. The Heaviside function is equal to unity (zero) if the value inside the brackets is positive (negative). X_i are the points on the reference trajectory and X_j are other points on the attractor in the vicinity of X_i. $|X_i - X_j|$ is the separation distance between the two points. $\theta(r - |X_i - X_j|)$ returns a value of unity if the distance between the two points is less than the hypersphere radius; otherwise a value of zero is returned. In this way, the Heaviside function 'counts' all the points within the hypersphere. The calculation of the correlation sum of equation (7.8a) involves following the reference trajectory, stopping at each discrete point on this trajectory and counting the number of other attractor points within a hypersphere of radius r. The cumulative sum of all the counted points is then divided by $N(N-1)$ to give the correlation sum, C_r. The maximum value of C_r is unity and this maximum is attained when the radius of the probing hypersphere is greater than the attractor's largest diameter and all points are counted. The minimum value of C_r is $2/[N(N-1)]$ and this is achieved when only the closest two points on the attractor are counted. Notice that the correlation sum counts the closest near neighbour twice (giving the two in the numerator) as the probing hypersphere visits both points on its journey around the attractor.

The correlation sum scales with the hypersphere radius according to a power law of the form

$$C_r \propto r^{D_C} \tag{7.8b}$$

where the exponent, D_C, is the correlation dimension. Hence, by examining the attractor in the method described above, for many different hypersphere radii, D_C is obtained from the slope of the **scaling region** of a $\log(r)$–$\log(C_r)$ plot, as shown in figure 7.3(b). The scaling region is the main linear region of the correlation curve, bounded by two nonlinear curved regions where large- and small-scale effects on the attractor influence the slope. In general, good estimates of D_C require large numbers of data points, N (see

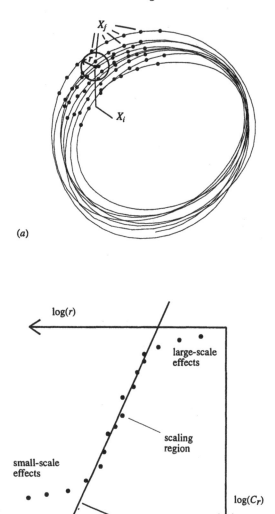

(a)

(b)

Figure 7.3. Determining the correlation dimension. (*a*) Probing hypersphere on the attractor. (*b*) The 'log(*r*)–log(*C_r*)' plot.

also section 7.8); however, the computational time required to calculate C_r increases with N^2. As is often the case in computational problems, a balance must be struck between accuracy and computing time.

The correlation dimension, information dimension and box counting dimension are related. In fact, they are part of a generalized collection of dimensions, known as Renyi dimensions. It can be shown that D_C forms a lower bound to D_I which in turn forms a

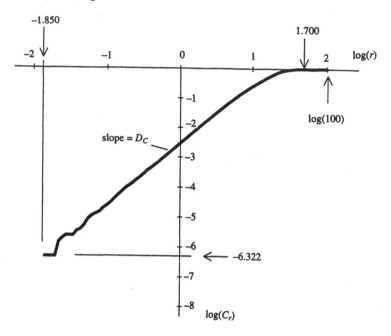

Figure 7.4. A correlation dimension plot for the Lorenz strange attractor.

lower bound to D_B,

$$D_C \leq D_I \leq D_B. \tag{7.9}$$

In addition, D_C and D_I asymptotically approach the value of D_B as the attracting set becomes more uniformly distributed in phase space. If the points on the attractor are uniformly distributed then $D_C = D_I = D_B$. In many practical cases D_C gives a very close estimate of D_B.

Figure 7.4 contains a $\log(r)$–$\log(C_r)$ plot for the Lorenz strange attractor. The attractor was generated using $N = 16\,384$ data points. To save computational time, the probing hypersphere stopped at every 128th data point on the attractor to measure the scaling of the attractor, i.e. only 128 ($= 16\,384/128$) locations were probed; we will call this number N_p. Hypersphere radii ranged from 0.01 to 100. The simplest way to choose this range is to use trial and error. However, a better way is to use the time series to make an educated guess for the maximum radius. If we refer back to the Lorenz time series of figure 6.8 we see that the ranges of the x, y and z variables are approximately 35, 50 and 45 respectively. The attractor should fit into a box of these dimensions. The largest radius required of a probing hypersphere should be of the order of the diagonal of this box ($= 75.8$). 100 was then chosen as the next-nearest round number for largest hypersphere radius. The correlation sum becomes one (i.e. $\log(C_r) = 0$) when all points on the attractor are within the hypersphere. This occurs at $\log(r) = 1.700$ in the plot, that is $r = 50.1$, which confirms our approximate value obtained from visual inspection of the time series. The minimum hypersphere radius range is more difficult to estimate, and is best selected by successively reducing the minimum radius until no neighbouring points are picked up within the hypersphere anywhere on the trajectory. In the plot of figure 7.4 this occurs at $\log(r) = -1.850$, that is $r = 0.014$. The correlation sum at this point is -6.322, which is $\log[1/(N_p(N-1))]$, corresponding to only one neighbouring point being located, divided by the maximum number of points that could be counted (equal to the number of points on the attractor visited multiplied by the maximum number of neighbouring points that may be counted, i.e. $128 \times 16\,383$). Notice that, in this case, the nearest neighbour has been counted only once. This is due to the probing hypersphere stopping at every 128th point, which has meant, in this case, that the correlation sum has missed out one of the nearest-neighbour counts.

As r approaches its maximum value, the $\log(r)$–$\log(C_r)$ curve begins to taper off, becoming horizontal as further increases in the hypersphere radius do not increase the correlation sum. As r approaches the minimum value, the $\log(r)$–$\log(C_r)$ line begins to fluctuate due to the small number of points contributing to the correlation sum. Between the maximum and minimum hypersphere radii, the $\log(r)$–$\log(C_r)$ curve becomes linear and the slope is equal to the correlation dimension, D_C. The slope in figure 7.4 is approximately 2.04, corresponding to the fractal interleaving of the essentially planar Lorenz attractor lobes.

7.5.4 The pointwise and averaged pointwise dimension

Another dimension estimate, closely related to the correlation dimension described above, is the **pointwise dimension**, D_P. Rather than looking at the scaling properties of the attractor as a whole we turn our attention to localized scaling of the attractor at a specific point on the attractor, say at point X_i. To do this we count only the number of neighbouring points, P_r, contained within the hypersphere at this location over a range of radii, r, more formally,

$$P_r = \frac{1}{(N-1)} \sum_{j=1;\,j\neq i}^{N} \theta \left(r - |X_i - X_j| \right). \qquad (7.10a)$$

P_r has a power law dependence on the radius r, given by

$$P_r \propto r^{D_P} \qquad (7.10b)$$

where the exponent, D_P, is the pointwise dimension. D_P may be computed at a specific location (point) on the attractor from a $\log(r)$–$\log(P_r)$ curve in a manner analogous to the correlation dimension. The pointwise dimension is a localized dimension estimate and in practice may vary over the entire attractor. To this end, an **averaged pointwise dimension**, D_A, may be defined as

$$D_A = \frac{1}{N} \sum_{i=1}^{N} D_P. \qquad (7.10c)$$

Although it has been suggested by some authors that this dimension estimate, D_A, is preferable to D_C as a measure of the average fractal properties of strange attractors, the correlation dimension is the more popular due to its computational speed and the wealth of background literature on its use.

7.5.5 The Lyapunov dimension

A relationship between the fractal dimension of strange attractors and their spectrum of Lyapunov exponents has been proposed. The Kaplan–Yorke conjecture, named after its proponents, defines the **Lyapunov dimension** as

$$D_L = j + \frac{\sum_{i=1}^{j} \lambda_i}{|\lambda_{j+1}|} \tag{7.11a}$$

where j is the largest integer to give

$$\sum_{i=1}^{j} \lambda_i > 0 \tag{7.11b}$$

and the spectrum is ordered in the usual fashion, $(\lambda_1, \lambda_2, \lambda_3, \ldots, \lambda_n)$, with $\lambda_1 > \lambda_2 > \lambda_3 > \ldots > \lambda_n$.

To clarify what we mean by equations (7.11a) and (7.11b), let us look briefly at a specific case of a chaotic attractor in 3D phase space. We know that for a chaotic attractor in 3D phase space there is zero divergence along the trajectory; divergence takes place in one orthogonal direction; and convergence in another orthogonal direction. These orthogonal directions are the principal axes of the ellipsoid of figure 7.2(d). If we centre an infinitesimally small cube, of side length ε, on the attractor trajectory, with its edges aligned in each of the principal directions, then after time t the cube will be deformed as shown in figure 7.5(a). In the direction of stretching (positive λ), the side length has stretched to $\varepsilon_{t(1)} = \varepsilon_0 \, e^{\lambda_1 t}$ with $\varepsilon_{t(1)} > \varepsilon_0$. Along the trajectory no stretching takes place, $\varepsilon_{t(2)} = \varepsilon_0 \, e^{\lambda_2 t}$, $\lambda_2 = 0$, hence $\varepsilon_{t(2)} = \varepsilon_0$. In the third orthogonal direction convergence occurs and λ_3 is negative, hence $\varepsilon_{t(3)} < \varepsilon_0$. The volume of the deformed cube is found simply by multiplying together the three side lengths, i.e.

$$V_d = \varepsilon_{t(1)} \varepsilon_{t(2)} \varepsilon_{t(3)} \tag{7.12a}$$

or

$$V_d = \varepsilon_0^3 \, e^{(\lambda_1 + \lambda_3)t} \tag{7.12b}$$

(remember $\lambda_2 = 0$). We would now like to find a relationship between the hypercube and its deformed self in terms of a dimension estimate. We can do this by covering both of them with hypercubes of side length δ. To make life easier, we will make δ equal to the smallest side of the deformed cube, $\varepsilon_{t(3)}$ (figure 7.5(b)). Our covering hypercubes then have volume

$$V_c = \delta^3 = \left(\varepsilon_0 \, e^{\lambda_3 t} \right)^3. \tag{7.13}$$

The number of these smaller hypercubes required to cover the deformed hypercube is denoted by N: it is found by dividing V_d by V_c, as follows:

$$N = \frac{\varepsilon_0^3 \, e^{(\lambda_1 + \lambda_3)t}}{(\varepsilon_0)^3 \, e^{3\lambda_3 t}} = e^{(\lambda_1 - 2\lambda_3)t} \tag{7.14}$$

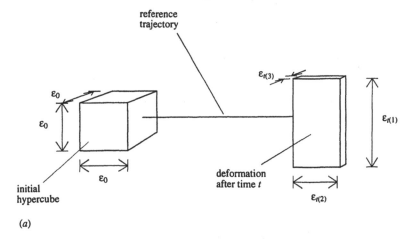

reference
trajectory

initial
hypercube

deformation
after time t

(a)

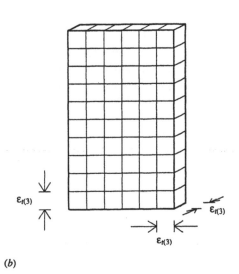

(b)

Figure 7.5. The Lyapunov dimension definition sketch. (a) The deformation through time of a hypercube centred on the reference trajectory. (b) Covering the deformed hypercube with elements of side length $\varepsilon_{t(3)}$.

where N is a function of time. As time increases, the initially small hypercube deforms for very large times and it approximately covers the entire attractor. In addition, the side length of the covering cubes, $\delta = \varepsilon_{t(3)} = \varepsilon_0 e^{\lambda_3 t}$, quickly decreases to zero. In the limit, as the time goes to infinity, we can approximate the box counting dimension of equation (7.6)) as follows:

$$D_L = \lim_{t \to \infty} \left[\frac{\log(e^{(\lambda_1 - 2\lambda_3)t})}{-\log\left(\varepsilon_0 e^{\lambda_3 t}\right)} \right] = \lim_{t \to \infty} \left[\frac{(\lambda_1 - 2\lambda_3) t}{-\log(\varepsilon_0) - \lambda_3 t} \right] \qquad (7.15a)$$

and, in the limit, this becomes

$$D_L = 2 - \frac{\lambda_1}{\lambda_3}. \qquad (7.15b)$$

Compare this with the more general expression of equation (7.11) (remember $\lambda_2 = 0$ and λ_3 is negative). Thus, the Lorenz attractor, with Lyapunov spectrum (2.16, 0.00, −32.4), has a Lyapunov dimension, D_L, of 2.07.

7.6   Attractor reconstruction

Experimental dynamical systems require special consideration as they generally exhibit highly complex behaviour, have many (even infinite) degrees of freedom and suffer from noise. To test an experimental system for chaos we first require an attractor. The problem is that typically only a single time series is available (usually this is a displacement— or velocity—time series). To construct an attractor from such an experimental time series we could construct a pseudo-phase space out of the x–$\dot{x}$–$\ddot{x}$–$\dddot{x}$–... co-ordinates. However, the resolution of the signal is not normally sufficiently accurate to allow the calculation of these higher order differentials. Thus, the **method of time delays** is often employed, where a reconstructed attractor (or pseudo-attractor) is embedded within an embedding space (or pseudo-phase space) of some suitable dimension, m (as distinct from the phase-space dimension, n, of the underlying dynamics).

In the method of time delays, we embed the attractor in an m-dimensional space and the m-dimensional co-ordinates, X_i, of the attractor are constructed from the original sampled time series, x_i ($i = 1, 2, 3, \ldots N$), as follows:

$$X_i = (x_i,\ x_{i+\xi},\ x_{i+2\xi},\ x_{i+3\xi},\ x_{i+4\xi}, \ldots x_{i+(m-1)\xi}) \qquad i = 1, 2, 3, \ldots N - (m-1)\xi$$

$$(7.16)$$

where X_i is the m-dimensional attractor vector produced from the discretely sampled time series x_i, and ξ is the discrete delay increment, or simply the delay. The time delay between reconstruction variables is therefore $\gamma = \xi dt$, where dt is the time interval between the ith point and the $(i+1)$th point on the time series, known as the sampling interval. Schematic diagrams of a time series and its reconstructed attractor are shown in figures 7.6(a) and (b) respectively. Figure 7.6(a) shows an arbitrary point 'j' on the time series (i.e. at $i = j$) and its two delay partners at $j + \xi$ and $j + 2\xi$. These three points are then used to construct the jth data point of a pseudo-attractor in 3D embedding space, $X_j = (x_j,\ x_{j+\xi},\ x_{j+2\xi})$. The point X_j is shown on the reconstructed attractor of figure 7.6(b) together with its two neighbouring points on the reconstructed trajectory at X_{j-1} and X_{j+1}. For example, say we want to construct a 3D attractor from the following time series, (1.01, 2.32, 1.93, 4.53, 2.22, 3.33, 3.34, 5.43, 5.55, 3.33, 4.02,...) with a delay increment $\xi = 3$. The first four co-ordinates of the reconstructed attractor would be $X_i = (x_i,\ x_{i+\xi},\ x_{i+2\xi})$; $i = 1, 2, 3, 4$, that is, (1.10, 4.53, 3.34), (2.32, 2.22, 5.43), (1.93, 3.33, 5.55) and (4.53, 3.34, 3.33).

The method of time delays provides a relatively simple way of constructing an attractor from a single experimental time series. However, one problem still remains; that is how to choose the delay, ξ. The choice of ξ is, in fact, non-trivial as we want the

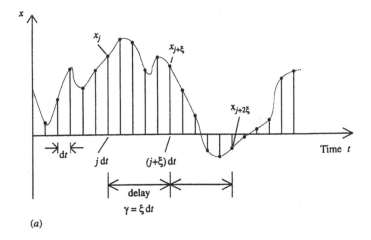

(a)

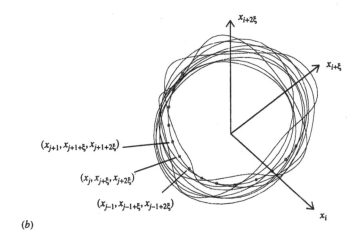

(b)

Figure 7.6. Attractor reconstruction using the method of time delays. (*a*) A discretized time series. (*b*) A schematic diagram of the reconstructed attractor. The *j*th point is shown together with its two neighbours on the trajectory.

dynamical properties of the reconstructed attractor to be amenable to subsequent analysis. In principle, if we have a very long and very accurate time series then any delay time should do. In practice, however, this is never the case, and we require to 'spread out' the reconstructed attractor in the embedding space so that it gives a good representation of the attractor geometry in phase space. This is especially true where small amounts of noise are present in the signal. Four common methods for determining an optimum delay are outlined below.

7.6.1   Method 1—visual inspection of reconstructed attractors

The simplest way of choosing ξ is to consider successively larger values of ξ and then visually inspect the phase portrait of the resulting attractor, choosing ξ which appears to give the most spread out attractor. This method has been used in practice but will only produce reasonable results with relatively simple systems.

7.6.2   Method 2—dominant period relationship

For low-dimensional attractors occurring in systems with forcing of a dominant period or a system with a dominant self-excited period, it is recommended that the most favourable value of the time delay is *one quarter of the dominant period*. This is a quick and easy method of determining ξ for such systems (knowing dt), however, many complex systems do not possess a single dominant forcing frequency.

7.6.3   Method 3—the autocorrelation function

The **autocorrelation function**, C, compares two data points in the time series separated by delay ξ, and is defined as,

$$C = \frac{\sum_{i=1}^{N-\xi} (x_i')(x_{i+\xi}')}{\sum_{i=1}^{N-\xi} (x_i')^2} \tag{7.17a}$$

where

$$x_i' = x_i - \overline{x_i} \tag{7.17b}$$

i.e. the fluctuating component of the time series, x_i', is separated from the temporal mean component, $\overline{x_i}$. The delay for attractor reconstruction, ξ, is then taken at a specific threshold value of C. Among the most popular of the many recommendations that have been made for this threshold are: the value of ξ which first gives C equal to one half; again the value of ξ which first gives C equal to zero; or the first inflection point of C. The inconsistent behaviour associated with the use of C to obtain a delay is illustrated in figure 7.7, where the autocorrelation function of both the x and z variables of the Lorenz strange attractor is plotted against the delay, ξ. Notice that the autocorrelation function exhibits a fluctuating decay with increasing time. The fluctuations are more marked for the z variable which does not exhibit the 'flipping' behaviour of the x variable (see figure 6.8). If we take the delay reconstruction time as that which gives $C = 0.5$, then $\xi = 12$ for the z variable and $\xi = 23$ for the x variable.

7.6.4   Method 4—the minimum mutual information criterion

It has been argued that, whereas the autocorrelation function measures the linear dependence of two variables, the mutual information function measures the general dependence of two variables. The mutual information of the attractor reconstruction co-ordinates (7.16) is defined as

$$M = \sum_{i=1}^{N-(m-1)\xi} P(x_i, x_{i+\xi}, x_{i+2\xi} \ldots x_{i+(m-1)\xi})$$

$$\times \log \left[\frac{P(x_i, x_{i+\xi}, x_{i+2\xi} \cdots x_{i+(m-1)\xi})}{P(x_i)P(x_{i+\xi})P(x_{i+2\xi}) \cdots P(x_{i+(m-1)\xi})} \right] \tag{7.18}$$

where $P(x_i)$ is the probability of occurrence of the time series variable x_i; and $P(x_i, x_{i+\xi}, x_{i+2\xi}, \ldots, x_{i+(m-1)\xi})$ is the joint probability of occurrence of the attractor co-ordinate $X_i = (x_i, x_{i+\xi}, x_{i+2\xi}, \ldots, x_{i+(m-1)\xi})$. M is a measure of the statistical dependence of the reconstruction variables on each other. If the co-ordinates are statistically independent, then

$$P(x_i, x_{i+\xi}, x_{i+2\xi} \cdots x_{i+(m-1)\xi}) = P(x_i)P(x_{i+\xi})P(x_{i+2\xi}) \ldots P(x_{i+(m-1)\xi}) \tag{7.19}$$

and it follows that $M = 0$. This would be the case for a completely stochastic (random) process, such as white noise. In contrast, complete dependence results in $M = \infty$. A suitable choice of time delay requires the mutual information to be a minimum. When this is the case the attractor is as 'spread out' as possible. This condition for choice of delay time is known as the **minimum mutual information criterion**.

The practical implementation of the minimum mutual information criterion is illustrated in a simple example in figure 7.8. A 2D reconstruction of an attractor is shown, in the $x_i, x_{i+\xi}$ plane (figure 7.8(a)). To calculate the mutual information, the plane is partitioned into a grid of N_c columns and N_r rows. Discrete probability density functions for x_i and $x_{i+\xi}$ are generated by simply summing the data points of the attractor lying in each column and row of the grid respectively and dividing by the total number of attractor points. $P(k)$ and $P(l)$ are the probabilities of occurrence of the attractor in column k and row l respectively. The joint probability of occurrence, $P(k, l)$, of the attractor in any particular box is calculated by counting the number of discrete points in the box and dividing by the total number of points on the attractor trajectory. The mutual information is found from

$$M = \sum_{k=1}^{N_c} \sum_{l=1}^{N_r} P(k, l) \log \left[\frac{P(k, l)}{P(k)P(l)} \right]. \tag{7.20}$$

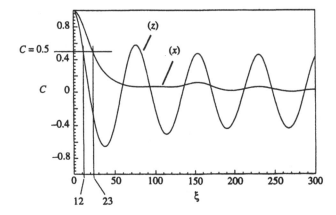

Figure 7.7. The autocorrelation function against delay for the x and z variables of the Lorenz strange attractor.

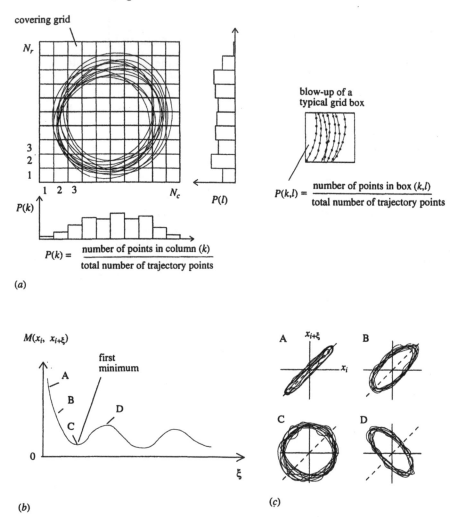

Figure 7.8. Mutual information. (*a*) Finding the mutual information. (*b*) Mutual information against reconstruction delay. (*c*) Typical attractor forms.

The value of ξ which gives the first minimum in the mutual information function is chosen as the attractor reconstruction delay. A typical plot of M against ξ is shown in figure 7.8(*b*). Typical forms for the attractor relating to the points A–D in figure 7.8(*b*) are sketched in figure 7.8(*c*). At very small delays, e.g. point A, the attractor co-ordinates are formed from very close points on the time series, hence there is a very strong positive correlation and the attractor aligns itself with the 45° line, the dashed line plotted in the figure. As the time series points used for the attractor co-ordinates begin to separate, and decorrelate, the attractor opens up in phase space (attractor D) until a minimum in M is reached (attractor C). For delays beyond point C, the mutual information increases and the attractor begins to close up in phase space (attractor D). It has been found that the first minimum in the mutual information of a 2D attractor reconstruction is generally adequate for determining a suitable delay for higher dimension reconstructions, in m-dimensional embedding spaces.

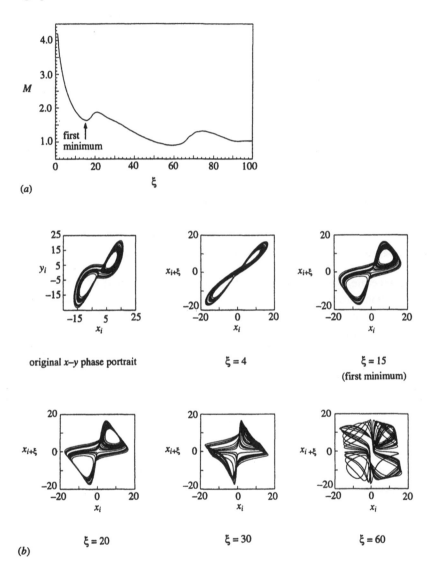

(a)

(b)

original x–y phase portrait

$\xi = 4$

$\xi = 15$
(first minimum)

$\xi = 20$

$\xi = 30$

$\xi = 60$

Figure 7.9. The minimum mutual information criterion and reconstruction of the Lorenz attractor. (a) Minimum mutual information against delay for the Lorenz equations—first minimum at $\xi = 15$ (d$t = 0.01$, hence the reconstruction time delay is 0.15.) (b) The original x–y phase portrait of the Lorenz attractor with five reconstructions using various delays.

Using the method described above, the mutual information is calculated for the Lorenz attractor and plotted against reconstruction delay in figure 7.9(a). The first minimum occurs at a delay, ξ, of 15. The original attractor is constructed in phase space using a non-dimensional time step, dt, of 0.01. The delay time is therefore, ξ d$t = 0.15$. Reconstructed attractors are shown in figure 7.9(b) together with an x–y phase portrait of the original strange attractor. At $\xi = 4$, the attractor is closed up and aligned with the 45° line. At $\xi = 15$, the first minimum in mutual information, the attractor has opened up. At $\xi = 20$, a local maximum in the mutual information occurs, corresponding to the tight compaction of the trajectories at each 'end' of the reconstructed attractor. As ξ increases further, the shape of the original Lorenz attractor becomes more difficult to discern, and by $\xi = 60$, the second minimum in mutual information, the attractor is quite unrecognizable.

Referring back to section 7.3 and our frequency spectra plots of figure 7.1(d), we found the non-dimensional orbital frequency of the Lorenz attractor to be approximately 1.35. This gives an orbital time of 0.74. If we consider this to be a dominant frequency of the system, then we could set the delay to one-quarter of this period, i.e. ξ d$t = 0.185$: a result not too different from the delay given by the minimum mutual information criterion, but producing an attractor which is significantly more 'closed up' in phase space (cf the $\xi = 15$ and 20 attractors of figure 7.9(b)).

7.7 The embedding dimension for attractor reconstruction

Now we can generate our reconstructed attractor using the method of time delays, with the time delay found from the mutual information criterion. We must do one more thing before we can characterize the attractor: that is, determine the dimension, m, of the embedding space required to adequately contain it. In other words, we must find the minimum embedding dimension which will allow it to possess the same geometrical properties as the original phase space attractor of the system. According to Whitney's embedding theorem a pseudo-attractor may, in general, be reconstructed within an embedding space of dimension

$$m = 2n + 1 \tag{7.21a}$$

where n is the dimension of the phase space of the underlying attractor. Alternatively, if the box counting dimension of the attractor is known then a closer estimate of the minimum embedding space required is

$$m > 2D_B \tag{7.21b}$$

(remembering m is an integer, although D_B need not be). For example, to reconstruct a pseudo-strange attractor from the chaotic Lorenz time series, knowing that it has a strange attractor in x–y–z phase space with a box counting dimension of just over two, we would require an embedding space of five to be sure of a reconstructed attractor with the same dynamical properties of the original attractor. If only D_C is to be measured then the requirement for an accurate dimension estimate is simply

$$m > D_C. \tag{7.21c}$$

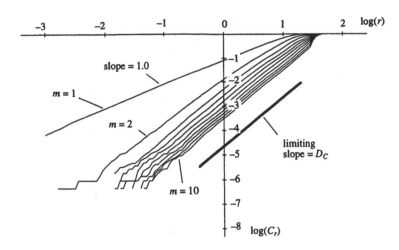

Figure 7.10. A correlation dimension plot for a reconstructed Lorenz strange attractor in embedding spaces of dimension $m = 1$–10.

One further problem arises with experimental signals, that is, we may not know the dimension of the underlying attractor or the phase space of the dynamical system. In this case, the simplest, practical method to determine a minimum dimension of embedding space is to calculate the correlation dimension for the reconstructed attractor in embedding spaces of successively larger dimensions. The slope of the $\log(r)$–$\log(C_r)$ plot will initially increase with the embedding dimension, reaching a limiting value when the embedding space is large enough for the attractor to untangle itself. Further increases in the embedding dimension should not increase the correlation dimension. This is illustrated in figure 7.10, where the correlation plots for the Lorenz strange attractor are given for embedding dimensions, m, from one to ten. Notice that the slope of the $\log(r)$–$\log(C_r)$ line is 1.0 for the attractor embedded in a 1D space, i.e. along a line. The maximum slope for the attractor embedded in ($m =$) 2D space is 2.0. We see that the slope of the lines in the figure soon reach a limiting value of $D_C \simeq 2.06$. This is in close agreement with the value of 2.04 we found above using the original attractor in 3D phase space. The limiting slope is drawn bold in the figure, beneath the $\log(r)$–$\log(C_r)$ correlation curves.

7.8 The effect of noise

Experimental attractors always contain a small random element known as noise. This may be due to either random fluctuations in the dynamical system—**dynamical noise**— or it may be due to random errors added to the signal by the measurement system— **measurement noise**. Noise fills up phase space densely, and causes the dimension of the resulting attractor to increase with, and be equal to, the embedding dimension, m. This is illustrated in figure 7.11. Figure 7.11(a) contains the phase portraits of pseudo-limit cycles which have been constructed using a sinusoidal function with noise levels increasing from 0 to 100% of the peak to peak amplitude. The noise is taken

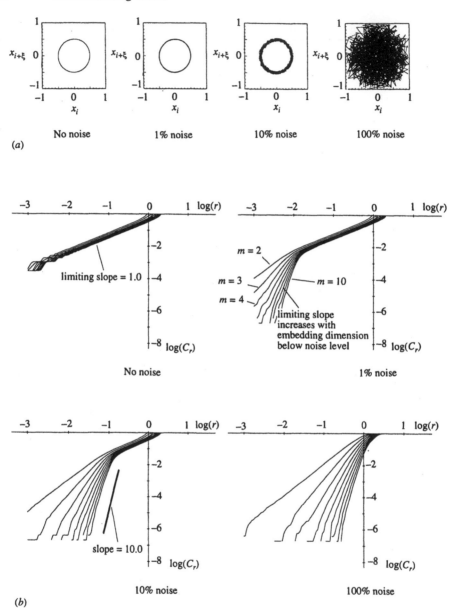

Figure 7.11. The effect of noise on correlation plots. (*a*) Attractor plots. (*b*) Correlation plots corresponding to the attractors in (*a*).

from a random number generator with a constant p.d.f. (see figure 4.5(a), left-hand diagram). In figure 7.11(b), the correlation plots for each of the limit cycles are given for embedding dimensions from two to ten. The limit cycle containing no noise results in a correlation plot where all the curves have slope equal to one, i.e. the correlation dimension, D_C, of the limit cycle curve. The correlation plot for the 1% noise level limit cycle contains two distinct parts. For probing radii, r, greater than the noise level the plot produces correlation curves of unit slope for $m = 2$–10, as did the case for 0% noise. However, below the noise level, the curves diverge and take up slopes of the order of the embedding dimension of the limit cycle. The 10% noise level shows the same behaviour: this time the curves diverge at larger values of r, taking up values equal to the embedding dimension. A line of gradient equal to ten is drawn in the figure for comparison. The 100% noise level completely obscures the structure of the limit cycle and correspondingly the correlation plot contains only curves with gradients equal to the embedding dimension. The correlation plots of figure 7.11(b) use base ten logarithms, hence the noise levels, indicated by the slope changes, jump up one unit (or one decade) on the $\log(r)$ axis for each multiple of ten of the noise level.

The effect of noise on the Lorenz attractor is illustrated in figure 7.12. Two attractors are plotted, one with 1% noise level, and one with 10% noise level. This time the correlation plots contain the correlation curves calculated from embedding dimensions from one to ten. As with the noisy limit cycles investigated above, the correlation plots for the noisy Lorenz attractors contain two distinct regions: one of uniform slopes for high embedding dimension, and a noisy region where the slopes increase along with increasing embedding dimension. However, this time the limiting slopes are affected noticeably by the noise level, giving D_C as approximately 2.16 and 2.35 for the 1% and 10% noise levels respectively.

Problems also occur when attempting to calculate the Lyapunov exponents from a noisy attractor. The trajectories become less well defined, due to the random fluctuations, and will have a probability distribution associated with them. After the trajectories have evolved for the elapsed time, t, there may be a small probability that the new trajectory points occur closer together, leading to a negative exponent. If the new points become very close, i.e. ε_t near zero, this will produce a very large negative Lyapunov exponent which will swamp the averaged value (see equations (7.2b) and (7.3)). The problem is shown schematically in figure 7.13. Worse still, if we are using a finite resolution then there is a possibility that the two new trajectory points will coincide (within the limits of resolution) at which point equation (7.2b) becomes undefined.

One further problem with dynamical noise in a real system with multiple attractors, is that it adds random fluctuations to the solution of the system. These random fluctuations may kick the trajectory approaching one attractor onto another trajectory approaching a different attractor, resulting in a completely different final solution for the system. Essentially, the solution jumps from one basin of attraction to another.

7.9 Regions of behaviour on the attractor and characterization limitations

When using characterization techniques such as the Lyapunov exponent or the correlation dimension, the investigator must be aware of scale and other effects, and how they relate to the results obtained. There are four main **regions of behaviour** on an attractor: these

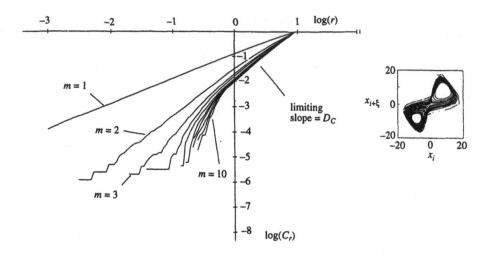

Level = 0.35 (approximately 1% of peak to peak amplitude of x-time series)

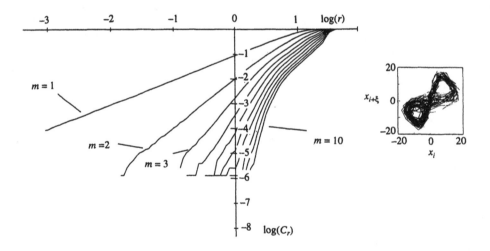

Level = 3.5 (approximately 10% of peak to peak amplitude of x-time series)

Figure 7.12. The correlation plots for noisy, strange Lorenz attractors.

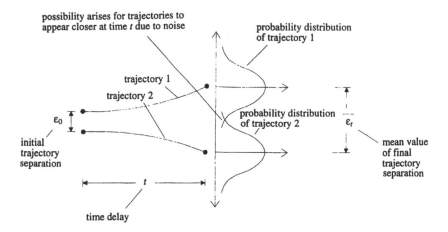

Figure 7.13. The effect of noise on the Lyapunov exponent calculation.

are outlined below, together with their effect on the correlation dimension estimate, beginning with the smallest length scales.

(i)   *Region 1.*   At very small scales the correlation dimension algorithm gives a dimension estimate of zero. This occurs when the radius of the probing n-dimensional hypersphere is less than the inter-point distances on the attractor trajectories (see figure 7.14(a)). Thus, the dimension estimate is that of a point, i.e. zero.

(ii)   *Region 2.*   The effect that this region has on the correlation dimension depends very much on whether we are investigating an experimental, and hence noisy, attractor, or a relatively noise free mathematical construction, such as the Lorenz attractor or the Rössler attractor.

    If the attractor is noisy, then for length scales of the order of the characteristic noise level (figure 7.14(b)), D_C will scale with the noise. That is, it will increase with, and should be equal to, the value of the embedding dimension, as we saw in figures 7.11 and 7.12.

    If, on the other hand, the attractor is noise free and the length scales being probed are of the order of consecutive points on the trajectory, then for limited data sets the algorithm will only detect points immediately nearby <u>on</u> the trajectory (figure 7.14(c)). This region of the attractor will show up as essentially 'linear' and the value of D_C will tend to unity. This region depends on whether consecutive points on the trajectory are closer than points on nearby trajectories. Thus, it is influenced both by the data sampling rate of the original time series and by the number of data points on it. One simple method which can be used to eliminate this effect is to discount from the correlation sum any point on the trajectory within one characteristic period of the motion.

(iii)   *Region 3.*   Once the hypersphere radius overcomes the effects of small length scales of regions 1 and 2, the hyper-sphere begins to probe the fractal structure of the attractor (figure 7.14(d)). The values of D_C approach those of the box counting

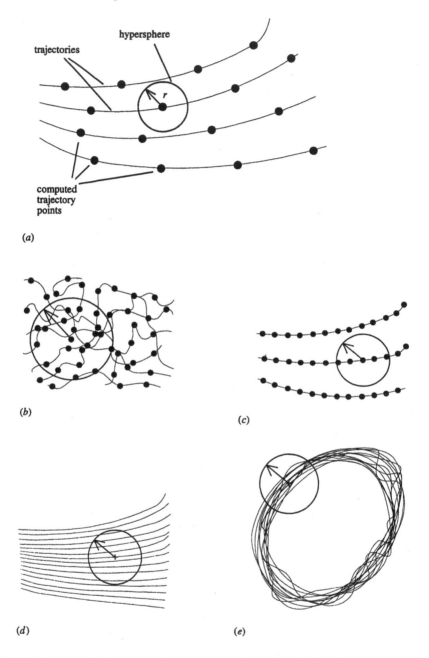

Figure 7.14. Regions of behaviour on the attractor. (*a*) Length scales less than inter-point distances on the attractor. (*b*) Short length scales on a noisy attractor. (*c*) Short length scales on a noise-free attractor. (*d*) Length scales leading to good dimension estimates. (Computed trajectory points are not shown for clarity.) (*e*) Length scales greater than the attractor radius. (Computed trajectory points are not shown for clarity.)

dimension, assuming that a suitable time delay has been chosen. This is the scaling region of the attractor (refer back to section 7.5 and figure 7.4).

(iv) *Region 4*. Once the length scales of the hypersphere are of the order of the attractor radius, as sketched in figure 7.14(*e*), edge effects dominate, and large parts of the hypersphere are outside the attractor and hence empty, thus reducing the value of D_C. At very large radii, the attractor begins to appear as a point within the hypersphere and the measured value of D_C tends, once again, to zero. We can see this 'tailing off' effect in figure 7.4 for large hypersphere radii.

In addition to the effects encountered in the regions of the attractor defined above, other properties of the attractor influence the measured value of the correlation dimension. The lacunarity of the attracting set has a bearing on the estimation of dimension (lacunar = gap or space). If, for example, an attractor is composed of distinct dense regions separated by gaps, i.e. it has a high degree of lacunarity, the value of D_C will fluctuate markedly as the probing hypersphere includes more and more of these distinct regions. Also, sparse data sets or limited data sets tend to produce errors in the calculations, which increase as the embedding dimension is increased. There are numerous proposals for the minimum number of data points required in order to give a good estimate of the correlation dimension. The Eckmann–Ruelle limit for the maximum correlation dimension that may be obtained from a time series of N points is

$$D_{C(MAX)} = \frac{2 \log(N)}{\log(1/\rho)} \tag{7.22}$$

where ρ is the maximum ratio of the radius of the probing hypersphere to the attractor radius which produces good dimension estimates, i.e. the maximum ratio encountered within the linear portion of the correlation curve. This ratio is often taken as 0.1, which assumes that by one decade of $\log(r)$ the slope will have become linear. As an example of the use of equation (7.22), the correlation dimension of the Lorenz strange attractor (figure 7.4) was found using an attractor of 16 384 data points. According to equation (7.22), this gives a maximum dimension of 8.4, well above the Lorenz attractor's fractal dimension of just over two. Another, more conservative, estimate by Smith states that the number of points required to estimate the correlation exponent of a nonlacunar attractor to within 5% of its true value increases at least as fast as

$$N_{min} = 42^j \tag{7.23}$$

where j is the largest integer that is less than the fractal dimension of the attractor. Thus, for the Lorenz attractor above, the minimum number of data points required to specify the attractor is $42^2 = 1764$ data points. It can be seen from equation (7.23) that the number of data points required for dimension estimates soon becomes computationally prohibitive. Other estimates have been proposed for the limits of the correlation dimension estimate, and details are given in the further reading section of this chapter. In addition, it should be noted that some investigators have found that acceptable results can be derived from data sets much smaller than those required by the above conditions.

The Eckmann–Ruelle limit for the number of data points required for a reliable estimate of the Lyapunov exponents of a chaotic system is given by

$$\log N = D_C \log(1/\rho). \tag{7.24}$$

If we compare this to equation (7.22), we see that in order to reliably estimate the Lyapunov exponents we need the square of the number of data points required to estimate dimension. Again, reasonably good estimates have been found, by various researchers, using fewer data points.

7.10   Chapter summary and further reading

7.10.1   Chapter keywords and key phrases

visual inspection	*broad band spectrum*	*Lyapunov exponent*
Lyapunov spectrum	*box counting dimension*	*information dimension*
correlation dimension	*scaling region*	*pointwise dimension*
averaged pointwise dimension	*autocorrelation function*	*Lyapunov dimension*
method of time delays	*measurement noise*	*dynamical noise*
minimum mutual information	*regions of behaviour*	
criterion	*(on the attractor)*	

7.10.2   General

The text by Hilborn (1994) covers many of the topics dealt with in this chapter in much greater (yet highly lucid) detail. Ott (1994) outlines recent developments in chaotic dynamics and, at a more advanced level, Ott *et al* (1994) provide a good collection of papers on the characterization of chaos together with a substantial explanatory text. An earlier collection of papers worth consulting is edited by Mayer-Kress (1986). Tufillaro *et al* (1992) give some preliminary details of the experimental techniques used in detecting chaos, and Bai-Lin (1990) edits a more advanced text devoted to the experimental study and characterization of chaos.

7.10.3   Fourier spectra

Newland (1993) and Ramirez (1985) provide good explanatory texts on the Fourier transform. There are many computer codes readily available for computing discrete Fourier transforms using the FFT method: see for example the work of Cooley *et al* (1969) or Press *et al* (1992). The paper by Farmer *et al* (1980) contains power spectra for the Lorenz and Rössler systems and, in addition, they relate the form of the spectra to the phase coherence of the chaotic system: see also Crutchfield *et al* (1980). Farmer (1981) describes the spectral broadening process of period doubling bifurcation processes. Appendix B contains power spectra for various experimental dynamical systems exhibiting chaos.

7.10.4   Dimension

Grassberger and Procaccia (1983a, b) suggested the use of the correlation dimension, D_C, as a practical algorithm for estimating the fractal dimension of a chaotic attractor. Grassberger (1983b) details the relationship between the correlation, information and box counting dimensions as part of a larger family of order-q Renyi dimensions, also known as generalized box counting dimensions, denoted D_q, where $-\infty < q < \infty$, and specifically $D_0 = D_B$, $D_1 = D_I$, $D_2 = D_C$. These generalized dimensions are useful in defining multifractal attractors which have spatially varying fractal properties. In addition, D_q is related to another measure of the multifractal property of an attractor, the multifractal scaling function, $f(\alpha)$: see for instance Hentschel and Procaccia (1983), Halsey *et al* (1986), Paladin and Vulpiani (1987), or, for a more gentle introduction, Hilborn (1994). A good review of many of the dimensions used in characterizing chaotic attractors is given by Farmer *et al* (1983). Hall and Wood (1993) discuss the performance of box counting dimension estimates, and Russell *et al* (1980) compare the box counting dimension and the Lyapunov dimensions for maps and ODEs. Grassberger (1990, 1993) and Hou *et al* (1990) present fast and memory box counting algorithms for computing the fractal dimension of data sets (computer program listings are included in the Grassberger papers). Moon and Li (1985b) have measured D_C for both experimental and numerical models of a Duffing oscillator (buckled beam). Beckman (1995) has produced an interesting paper on the visualization of point distributions in high dimensional spaces. Grassberger (1986, 1987) highlights the limitations in the correlation dimension estimate when attempting to locate an attractor for the **climate**. Farmer (1982) uses both a dimension estimate and Lyapunov exponents to characterize infinite dimensional systems in the form of the Mackey–Glass delay differential equation. Komori *et al* (1994) use both the correlation dimension and the largest Lyapunov exponent to characterize chaos in a **turbulent plasma**. Pottinger *et al* (1992) illustrate a visualization technique for attractors in phase spaces of dimension greater than three.

7.10.5 Lyapunov exponents

A search for information on the practical measurement of Lyapunov exponents should begin with the paper by Wolf *et al* (1985) in which program listings are given for the determination of the principal Lyapunov exponent, λ_1, and also $\lambda_1 + \lambda_2$. In addition, Lyapunov spectra are given for various systems including the Lorenz and Rössler systems together with their calculated Lyapunov dimensions (Kaplan and Yorke 1979). Methods for determining the Lyapunov spectra of a chaotic time series are given by Froyland and Alfsen (1984), Sano and Sawada (1985), Eckmann *et al* (1986) and Darbyshire and Broomhead (1996). In addition, Froyland and Alfsen (1984) present the Lyapunov exponents for the Lorenz model over a wide range of the r control parameter. Similarly, Zeni and Gallas (1995) have used the Lyapunov exponent to investigate the occurrence of chaos in the Duffing oscillator over a wide range of control parameters. The problem of noisy and spurious Lyapunov exponents is dealt with by Bryant *et al* (1990) and Parlitz (1992). Velocity dependent Lyapunov exponents have been developed by Deissler and Kaneko (1987) for characterizing chaotic systems where the dynamical processes of interest are in a moving frame of reference. Other papers of interest concerning Lyapunov exponents include those by Sri Namachchivaya *et al* (1994), Yang *et al* (1996) and Mehra and Ramaswamy (1996).

There are many examples in the literature of the use of dimension estimates and Lyapunov exponents to characterize dynamical systems, both real and modelled. A good example is by Brandstäter *et al* (1983), who use both characterization techniques to investigate chaos in **Taylor–Couette flow**. Brandstäter and Swinney (1987) investigate the same system and provide examples of the use the minimum mutual information criterion, power spectra, phase portraits and dimension estimates, together with a discussion of the potential problems associated with dimension estimates.

7.10.6 Embedding

Reconstructing attractors in an embedding space using a single time series was first suggested by Packard *et al* (1980) and proved theoretically as a legitimate technique by Takens (1981). Ding *et al* (1993) show that the correlation sum will produce consistent results for delay reconstructed attractors of long enough data sets when the number of delay co-ordinates exceeds the correlation dimension of the reconstructed attractor. Often, the required embedding dimension for attractor reconstruction is less than that specified by Whitney's embedding theorem (Whitney 1936), which relates to smooth manifolds. The minimum embedding dimension required based on the box counting dimension of the attractor is given by Sauer *et al* (1991). Kennel *et al* (1992) provide an additional criterion for determining the minimum required embedding dimension for attractor reconstruction based on the method of false nearest neighbours.

7.10.7 Delay reconstruction

Fraser and Swinney (1986) suggested the minimum mutual information criterion as a method determining the most appropriate delay, ξ, to use in the construction of a pseudo-attractor from a single time series. Broomhead and King (1986) have suggested another procedure to estimate the delay, ξ, based on singular valued decomposition (SVD): see also Albano *et al* (1988) who also detail the effect of the overall delay window, $(n-1)\xi$, on the correlation dimension estimate of reconstructed attractors. Fraser (1989) compares the minimum mutual information and SVD techniques and recommends the minimum mutual information criterion over the SVD method. However, there is still no general agreement over which method is best as the SVD technique has a built in noise reduction—very useful for noisy experimental attractors—but it measures only linear dependence in the reconstruction co-ordinates rather than the general dependence of the minimum mutual information criterion. There are many other methods suggested for choosing ξ (see Rosenstein *et al* (1994) for a discussion). Of course, if the time signal under investigation is of a high enough quality then an attractor may be reconstructed using the higher order derivatives of the signal, as Roux *et al* (1980) have done for a strange attractor in chemical turbulence, and, if you are a very lucky experimentalist with access to multiple simultaneous measurements of the dynamical system, then these can be used directly to form an attractor without employing the method of time delays (see for example Guckenheimer and Buzyna (1983)). See also the article by Buzug *et al* (1994) who detail a **Taylor–Couette experiment** with simultaneous time series measurement and a information dimension based upon the mutual information function. Other papers worth consulting on the question of attractor reconstruction are those by Buzug *et al* (1990), Mancho *et al* (1996), and Lai *et al* (1996).

7.10.8 Noise

Moss and McClintock (1989) provide a useful general collection of articles on noise in experimental and simulated nonlinear dynamical systems. The use of the correlation dimension on noisy attractors is discussed by Ben-Mizrachi *et al* (1984). Kapitaniak (1988b) introduces random Lyapunov exponents to determine the divergence properties of the attractors of a **noisy mechanical system** with stress relaxation. For more detailed information on the effect of noise and methods for noise reduction see initially the articles by Kostelich and Yorke (1990), Fraedrich and Wang (1993), Szpiro (1993), Grassberger *et al* (1993) and Pierson and Moss (1995). Low pass filtering of a measured time series signal is often performed to reduce noise. However, the filtering process itself can lead to problems in the measured dimension and Lyapunov exponents of the system (see for example Mitschke *et al* (1988), Badii *et al* (1988), Chennaoui *et al* (1990) and Broomhead *et al* (1992)). See also the article by Rosenstein and Collins (1994), who provide a method for visualizing the effects of filtering on chaotic data.

7.10.9 Limits

The data requirements for reliable dimension estimates remains an open question in the literature and there have been many limits suggested for the calculation of the correlation and other dimensions by various authors. General accounts of the data requirements for reliable dimension estimates are given by Albano *et al* (1987) and Mayer-Kress (1987). Limits for dimension and Lyapunov exponent calculations are outlined in detail in the papers by Eckmann and Ruelle (1992) and Ruelle (1990). Smith (1988) and Procaccia (1988) provide more conservative estimates. Theiler (1986) gives cause for concern when using small data sets, whereas Abraham *et al* (1986) report that relatively short data sets may provide good dimension estimates, and Gershenfeld (1992) reports that for certain cases dimensions in excess of ten can be reliably measured from laboratory data. Malinetskii *et al* (1993a, b) discuss the additional data sometimes required for a good dimension estimate from a reconstructed attractor rather than an attractor in original phase space. Dvorák and Klaschka (1990) provide a modification of the Grassberger–Procaccia correlation dimension algorithm which compensates for the edge effects on the attractor. Rosenstein *et al* (1993) give details of a method for calculating the largest Lyapunov exponent from small data sets.

7.10.10 Advanced topics

This chapter has described various methods for quantifying chaos. Dimension estimates reveal to us the geometrical complexity of the dynamics and the Lyapunov exponents, the divergence of errors, and hence prediction horizons of the system. A characteristic measure of chaos not mentioned in the chapter is the Kolmogorov entropy, related to Shannon's formula. It measures the change in entropy of the signal and is in fact equal to the sum of the positive Lyapunov exponents for the system. For more information see the work by Ott (1993), Cohen and Procaccia (1985) and Grassberger and Procaccia (1983c). Theiler *et al* (1992) have described a method for identifying nonlinearity in time series by comparing the original time series with a surrogate time series generated by randomizing the phases of the Fourier components of the original time series. Kaplan and Glass (1992) describe another method for testing for determinism in time series, that is, differentiating it from a purely random time series.

In many cases, we have no detailed knowledge of the underlying dynamics of a particular system being studied. If we have an experimental system exhibiting low dimensional chaos, we would like to be able to 'work backwards' and construct a dynamical model of it based upon its time series signal. Attempts have been made to do this, see for example the articles by Eckmann and Ruelle (1985) and Casdagli (1989). See Sugihara and May (1990) for an implementation of the ideas in experimental time series of **measles, chickenpox** and **marine phytoplankton populations**. Ott, Grebogi and Yorke (1990) have proposed a method (the OGY method) to control chaotic oscillations. This is done by adding weak perturbations to the dynamical system to force it to follow one of the unstable periodic orbits embedded within the chaotic attractor. There are now many control algorithms and their use is illustrated for the Lorenz system by Liu and Leite (1994), and for experimental systems by Ditto *et al* (1990a), Singer *et al* (1991), Azevedo and Rezende (1991), Auerbach *et al* (1992), Roy *et al* (1992) and Christini and Collins (1996). Lai (1996) describes a 'small' control method and alternative control methods are detailed by Rajasekar and Lakshmanan (1993).

We have used the terms strange attractor and chaotic attractor interchangeably throughout the text, as is the norm in the literature. However, some authors make the distinction between strange attractors, which have a fractal structure, and chaotic attractors, which exhibit divergence of nearby trajectories. Indeed, for special cases it is possible to have dynamical systems with strange, nonchaotic attractors, i.e. possessing a fractal structure and no positive Lyapunov exponents. Romeiras *et al* (1987) detail strange nonchaotic attractors for a quasiperiodically forced continuous dynamical system, and in maps by Grebogi *et al* (1984) and Heagy and Hammel (1994). Ditto *et al* (1990b) have found such attractors on a periodically driven **magnetoelastic ribbon experiment**.

7.11   Revision questions and further tasks

Q7.1 List the keywords and key phrases of this chapter and briefly jot down your understanding of them.

Q7.2 What difficulties arise in visually inspecting time series data to deduce whether a dynamical system is behaving chaotically?

Q7.3 Using sketches, explain the main differences between the frequency spectra associated with periodic and chaotic time series from a deterministic dynamical system. What is the limitation in using only frequency spectra to characterize experimental data?

Q7.4 It is known that a dynamical system has a phase space of dimension 3.

(*a*) If only one time series signal is available, what is the minimum dimension of the embedding space required to generate a reconstructed attractor with the same dynamical properties as the original phase space attractor?

(*b*) What theorem does (*a*) relate to?

(*c*) If the reconstructed attractor of (*a*) is known to have a box counting dimension, D_B, of 2.12, what revised estimate can be made of the minimum dimension of phase space required to embed the reconstructed attractor?

(*d*) If the reconstructed attractor of (*a*) is known to have a correlation dimension, D_C, equal to its box counting dimension, what further revised estimate can be made of the minimum dimension of phase space required to embed the reconstructed attractor?

(*e*) What practical method could you employ to find a minimum embedding dimension of a strange attractor formed from a time series if you did not know the dimension of the original phase space, n, the box counting dimension, D_B, or the correlation dimension, D_C?

Q7.5 The attractor of a dynamical system has the following spectra of Lyapunov exponents for two settings of the system control parameters: (2.813, 0, −9.543, −23.241) and (0, −9.314, −44.219, −48.894).

(a) Which of the two attractors is strange?—Explain why.

(b) Why do both spectra contain zero values of λ?

(c) Symbolically define the attractors.

Q7.6 The Lyapunov exponent is often defined, in terms of base two logarithms, as, $\lambda = \frac{1}{t} \log_2 [\varepsilon_t / \varepsilon_0]$. In this form the exponent measures the information loss of the system in bits per second. This is useful, as the measurement of real dynamical systems is often given in terms of bits of information (of resolution). Use this form of the Lyapunov exponent to answer the following.

An attractor is tested for chaos. The initial separation of two adjacent trajectories on the attractor is 0.0032 units in phase space. After an elapsed time of 0.005 s the two trajectories are now 0.0057 units apart.

(a) If these trajectories remain in the plane of maximum divergence of the attractor determine the Lyapunov exponent of the system in bits (of information loss) per second.

(b) From your answer to (a) state whether the system is chaotic or periodic.

Q7.7 Data points on the attractor of Q7.6 are measured with an eight bit resolution. If the orbital frequency of the trajectory on the attractor is 500 Hz, calculate the elapsed time and the number of orbits of the attractor after which current predictions would become invalid.

Q7.8 The spatial extent of the attractor in Q7.6 and Q7.7 is 0.1304 units in embedding space and we can now measure it to 0.0001 ± 0.00005 units. Assuming that we know the equations of motion of the dynamical system, what is the prediction horizon for the attractor, over which predictions based on current measurements would become invalid.

Q7.9 (a) What is the Lyapunov dimension of the chaotic attractor in Q7.5?

(b) What is the Lyapunov dimension of the Rössler hyperchaotic system with Lyapunov spectrum given in section 7.4?

Q7.10 Figure Q.7.10 gives a correlation plot for an attractor of a model journal bearing system in chaotic mode. Estimate the fractal dimension of the attractor from the plot. The scaling region is defined between the two dots on each correlation curve.

Q7.11 The first twenty points of a time series obtained from an experimental dynamical system are: 0.0, 1.1, 2.3, 3.3, 4.5, 6.4, 6.2, 5.8, 4.6, 3.3, 1.9, 1.7, 0.1, −0.3, −2.4, −3.9, −4.7, −4.1, −3.0, −1.2, ...

(a) Using a delay $\xi = 4$ generate the first fifteen points of the reconstructed attractor in a 2D phase space.

(b) Plot out the phase portrait of the attractor from the points obtained in (a).

(c) Using a delay $\xi = 3$ generate the first five points of the attractor in 3D phase space.

Q7.12 An investigator has used 10 000 data points to reconstruct a strange attractor of unknown dynamics in order to test it for chaos.

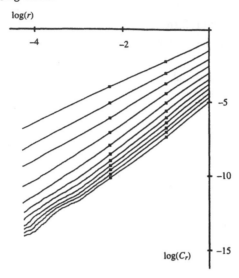

Figure Q7.10. Correlation plot for a model journal bearing system exhibiting chaos.

(*a*) What is the Eckmann–Ruelle limit on the correlation dimension that can be reliably calculated?

(*b*) What other factors affect the accuracy of the computation of D_C?

Q7.13 Write a computer program to estimate D_C (based on equations (7.8*a*) and (7.8*b*)) of the Lorenz attractor output from the program GENERAL in appendix A.

Q7.14 Using the output from the GENERAL program in appendix A, produce phase portraits of reconstructed Lorenz attractors for delays, $\xi = 4, 15, 20, 30,$ and 60. Compare your results with figure 7.9.

Q7.15 Write a program to estimate the minimum mutual information of Lorenz attractors reconstructed in 2D embedding space. (Base the program on equation (7.20).) Use the program to determine the first minimum in the mutual information, and hence suggest an optimal time delay for reconstruction of the Lorenz attractor.

Q7.16 Write a program to produce reconstructed attractors in embedding spaces of dimension, $2 \leq m \leq 10$. Using the optimal delay increment obtained in Q7.15 estimate D_C from reconstructed Lorenz attractors using the x–time series output from the program GENERAL in appendix A.

Q7.17 Repeat Q7.13 to Q7.16 using the Rössler attractor output from the program GENERAL. (See appendix A for required modification.)

Q7.18 Repeat Q7.16, this time add various levels of noise to the time series signal and investigate its effect on the correlation plot, and hence D_C. (Noise can be easily added by using the random number generator in your computer, which usually has a constant distribution. Add a scaled random number to each time series data point, repeat for the time series with various scalings and inspect the correlation plots. If possible, investigate the effect of other noise distributions on your results.)

Appendix A

Computer program for Lorenz equations

The FORTRAN program GENERAL listed below produces the time series output for the Lorenz equations using a fourth-order Runge–Kutta routine (kindly provided by Andrew Chan (1996)). The time series is output to the file GEN.DAT. In the output file the time, t_i, and the x_i, y_i and z_i co-ordinates of the Lorenz attractor are listed sequentially, where $i = 0$–NSTEP. Every OUTSTEP values are written to the data file. The initial conditions X0, Y0, and Z0 lie close to the attractor. NSTART allows the user to output post-transient data. The program is easily modified for other nonlinear equations of motion; the modification required for GENERAL to produce time series data for the Rössler equations is given below the main listing.

```
      PROGRAM GENERAL
C..   THIS PROGRAM WILL CALCULATE THE SOLUTION TO THE LORENZ
C..   EQUATIONS USING A 4TH ORDER RUNGE-KUTTA METHOD
      IMPLICIT NONE
      EXTERNAL EQNS
      INTEGER N,NSTEP,I,J,NSTART,OUTSTEP
      PARAMETER(N=3)
C..   N IS THE NUMBER OF FIRST ORDER O.D.E.'S
      REAL T0,DT,X0,Y0,Z0,TA,TB,YA(N),RK(N,4),Y(N),Y1(N),YB(N)
      REAL XP(0:30000),YP(0:30000),ZP(0:30000),T(0:30000)
      REAL SIGMA,B,R
      COMMON /PAR/ SIGMA,B,R
C
      PRINT *,'ENTER SIGMA'
      READ(6,*) SIGMA
      PRINT *,'ENTER B'
      READ(6,*) B
      PRINT *,'ENTER R'
      READ(6,*) R
C
      T0    =   0
      X0    =   3.0
      Y0    =   6.0
      Z0    =   8.0
```

```
      NSTART  =    0
      OUTSTEP =    1
      NSTEP   =    10000
      DT      =    0.01
      YA(1)=X0
      YA(2)=Y0
      YA(3)=Z0
      XP(0)=X0
      YP(0)=Y0
      ZP(0)=Z0
C
      J=0
      DO 10 I=1,NSTEP
         J = J + 1
         TA=T0+REAL(J-1)*DT
         TB=T0+REAL(J  )*DT
         CALL RUNGE(EQNS,TA,TB,1,YA,N,RK,Y,Y1,YB)
         YA(1)=YB(1)
         YA(2)=YB(2)
         YA(3)=YB(3)
C
C..   TIME SERIES COORDINATES FOR OUTPUT FILE
C
         XP(I)=YA(1)
         YP(I)=YA(2)
         ZP(I)=YA(3)
         T(I) = TB
10    CONTINUE
C
C..   WRITE TO DATA FILE
C
      OPEN(21,FILE='GEN.DAT')
      WRITE(21,101) (T(I),XP(I),YP(I),ZP(I),I=NSTART,NSTEP,OUTSTEP)
101   FORMAT(4F14.8)
      STOP 'COMPILATION FINISHED SUCCESSFULLY'
      END
C********************************************************************
      SUBROUTINE RUNGE(EQNS,TA,TB,NSTEP,YA,N,RK,Y,Y1,YB)
C..   SUBROUTINE FOR RUNGE KUTTA INTEGRATION (4TH ORDER)
      IMPLICIT NONE
      INTEGER NSTEP,N,I,J
      REAL TA,TB,YA(N),YB(N),Y(N),RK(N,4),H,Y1(N),T
      EXTERNAL EQNS
C
      H=(TB-TA)/REAL(NSTEP)
      DO 10 J=1,N
         Y(J)=YA(J)
   10 CONTINUE
      DO 100 I=1,NSTEP
         T=TA+REAL(I-1)*H
```

```
      CALL EQNS(T,Y,YB,N)
      DO 20 J=1,N
         RK(J,1)=YB(J)*H
         Y1(J)=Y(J)+RK(J,1)*0.5
  20     CONTINUE
      CALL EQNS(T+0.5*H,Y1,YB,N)
      DO 30 J=1,N
         RK(J,2)=YB(J)*H
         Y1(J)=Y(J)+RK(J,2)*0.5
  30     CONTINUE
      CALL EQNS(T+0.5*H,Y1,YB,N)
      DO 40 J=1,N
         RK(J,3)=YB(J)*H
         Y1(J)=Y(J)+RK(J,3)
  40     CONTINUE
      CALL EQNS(T+H,Y1,YB,N)
      DO 50 J=1,N
         RK(J,4)=YB(J)*H
         Y(J)=Y(J)+(RK(J,1)+RK(J,4)+2.0*(RK(J,2)
                  +RK(J,3)))/6.0
  50     CONTINUE
 100 CONTINUE
      DO 110 J=1,N
         YB(J)=Y(J)
 110 CONTINUE
C
      RETURN
      END
C******************************************************************
      SUBROUTINE EQNS(T,YI,YO,N)
C
C..   THE LORENZ EQUATIONS
C
      IMPLICIT NONE
      INTEGER N
      REAL T,YI(N),YO(N)
      REAL SIGMA,B,R
      COMMON /PAR/ SIGMA,B,R
C
      YO(1)=-SIGMA*(YI(1)-YI(2))
      YO(2)=-YI(1)*YI(3) + R*YI(1)-YI(2)
      YO(3)=YI(1)*YI(2) - B*YI(3)
C
      RETURN
      END
```

A.1   The Rössler modification

To generate a numerical solution to the Rössler equations using the program GENERAL substitute the following lines in the main program:

```
REAL A,B,C
COMMON /PAR/ A,B,C


PRINT *,'ENTER A'
READ(6,*) A
PRINT *,'ENTER B'
READ(6,*) B
PRINT *,'ENTER C'
READ(6,*) C


XO      =    0.0
YO      =   -3.6
ZO      =    0.0


DT      =    0.05
```

and replace the subroutine EQNS with the following:

```
      SUBROUTINE EQNS(T,YI,YO,N)
C
C..   THE ROSSLER EQUATIONS
C
      IMPLICIT NONE
      INTEGER N
      REAL T,YI(N),YO(N)
      REAL A,B,C
      COMMON /PAR/ A,B,C
C
      YO(1)=-YI(2)-YI(3)
      YO(2)=YI(1)+A*YI(2)
      YO(3)=B+YI(1)*YI(3)-C*YI(3)
C
      RETURN
      END
```

Appendix B

Experimental chaos

Mathematical models have been used extensively within the main text to illustrate various properties of chaos. However, chaotic behaviour has now been observed in nature over a broad range of scientific and engineering disciplines. In this appendix, a variety of experimental observations are reprinted together with a very brief explanatory text. The aim here is really to convince the reader of the universal nature of chaos, rather than give a detailed discussion of each case. A reference is provided for each system featured in this appendix. In addition, the further reading sections at the end of chapters 5–7 provide many more references to experimental observations of chaos.

B.1 The buckled beam

The magnetoelastic beam experiment of Moon and Holmes (1979) is shown in figure B.1(a). It consists of a slender steel beam suspended from a cantilever on a rigid frame. Two magnets attached to the base of the frame attract the beam. When the frame is at rest the beam is attracted to one or other of the magnets. The frame is oscillated periodically in the horizontal direction and the motion of the steel beam monitored through a strain gauge attached to its surface. Moon and Holmes found that the motion of the beam may be modelled using a modified Duffing equation of the form

$$\ddot{x} + r\dot{x} - \tfrac{1}{2}\left(x - x^3\right) = A_f \cos \omega t. \tag{B.1}$$

The top time series in figure B.1(b) shows chaotic oscillations in the output of the strain gauge, and the lower time trace shows a computer simulation of the motion, using the above equation. The parameter values used are given below each plot. A Poincaré map for another experimental chaotic case is plotted in B.1(c). The experiment provides a link between chaos in a real mechanical system and its numerical model.

B.2 The journal bearing

There has been much interest in the potential application of chaos theory to problems involving rotating machinery.   One engineering component which has received considerable interest is the journal bearing. The top of figure B.2 contains a schematic

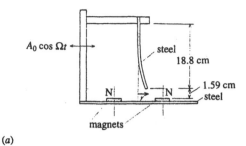

(a)

Experimental ($r = 0.0036$; $A_f = 0.035$; $\omega = 0.89$)

time

time

(b)   Modelled ($r = 0.045$; $A_f = 0.28$; $\omega = 0.84$)

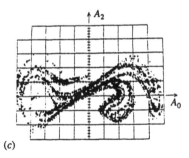

(c)

Figure B.1. Chaos in forced magnetoelastic beam buckling. (*a*) An apparatus sketch. (*b*) A chaotic time series output. (*c*) An experimental Poincaré map. ($r = 0.026$; forcing frequency $= 8.5$ Hz; forcing amplitude $= 3$ mm peak to peak.) (After Moon and Holmes (1979). Reproduced by permission of Academic Press.)

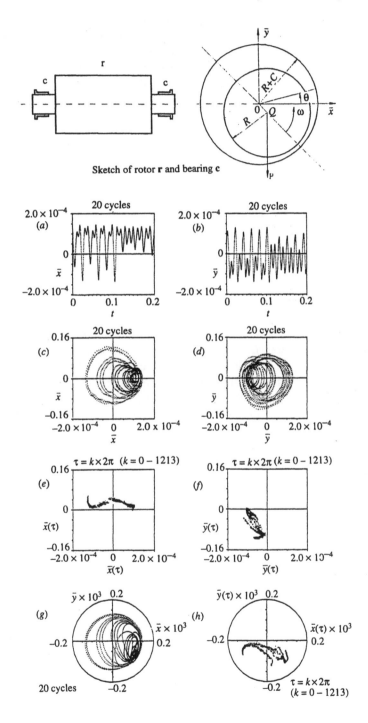

Figure B.2. Chaos in a journal bearing. (After Adiletta *et al* (1996). Reproduced with kind permission from Kluwer Academic Publishers.)

diagram of a rotor, **r**, supported either end by two journal bearings, **c**. The cross section of a journal bearing is shown on the right of the figure, where the rotor shaft rotates within a fluid filled gap between itself and the bearing casing. At high rotation rates the shaft may become unstable and produce chaotic oscillations (see also Brown *et al* (1994)). Figure B.2 also contains various forms of output for experimental chaos in a journal (labelled (*a*)–(*h*) in the figure). This behaviour was observed by Adiletta *et al* (1996a) for a bearing rotating at 6100 r.p.m. (*a*) and (*b*) show respectively the horizontal and vertical displacement of the rotating shaft centre against time. (*c*) and (*d*) plot velocity–displacement phase portraits of the motion for the horizontal and vertical co-ordinate axes respectively; and their corresponding Poincaré maps are shown in (*e*) and (*f*) respectively. Finally, (*g*) shows the actual motion of the shaft centre in *x–y* space, with its Poincaré map given in (*h*). The emergence of chaotic vibrations in the bearing can reduce bearing efficiency and lead to severe vibrational problems within the bearing supports.

B.3 Chua's circuit

Chua's circuit has allowed in-depth investigation using numerical, mathematical and experimental approaches (Madan 1993). In particular, computer models of the circuit have produced results extremely close to experimental measurements. The circuit (figure B.3(*a*)) contains an inductor, L, two capacitors, C_1 and C_2, a linear resistor R, and a nonlinear resistor N_R. This nonlinear circuit exhibits both periodic and chaotic oscillations. Two types of strange attractor have been observed in the circuit: a Rössler type attractor and one with a double scroll structure. Figure B.3(*b*) contains plots of attractors and power spectra from an experiment by Elgar and Kennedy (Madan 1993, pp 892–907). In the figure, a periodic attractor and both types of strange attractor are plotted, together with their associated power spectra. The broad-band nature of the chaotic spectra may be contrasted with the dominant nature of the spectral peaks which occur for the periodic case.

B.4 The Belousov–Zhabotinsky chemical reaction

Figure B.4 contains an example of chemical chaos for the Belousov–Zhabotinsky reaction taken from the paper by Roux *et al* (1983). Without going into too much detail, this chemical reaction produces temporal oscillations in the concentrations of the reacting chemicals. Figure B.4(*a*) plots the concentration of the time series of the bromide ion potential for both a periodic and chaotic case; their associated power spectra are given in figure B.4(*b*). Figures B.4(*c*) contains a limit cycle attractor for the periodic case. Finally, figure B.4(*d*) contains a set of reconstructed attractors, plotted at various reconstruction delays, showing a strange attractor for the chaotic case.

B.5 Taylor–Couette flow

Figure B.5 contains the time series and power spectra for a Taylor–Couette experiment carried out by Gollub and Swinney (1975). (Refer back to chapter 6, section 6.7, for more

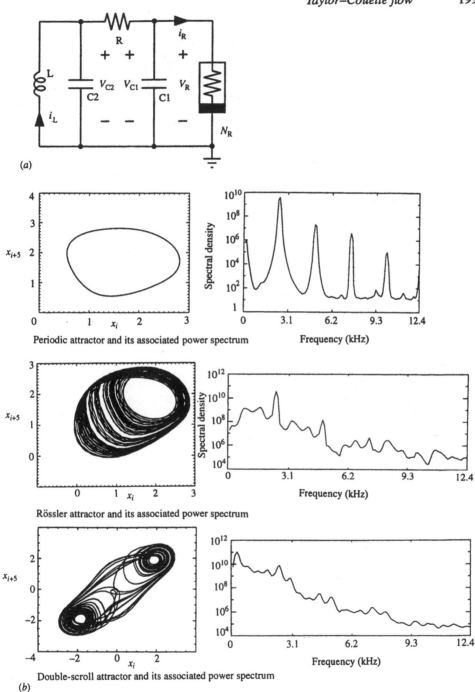

Figure B.3. Chaos in Chua's electric circuit. (*a*) Chua's circuit. (*b*) Three reconstructed attractors and their corresponding power spectra. (After Elgar and Kennedy (Madan 1993, pp 892–907). Reproduced by permission of World Scientific Publishing Co.)

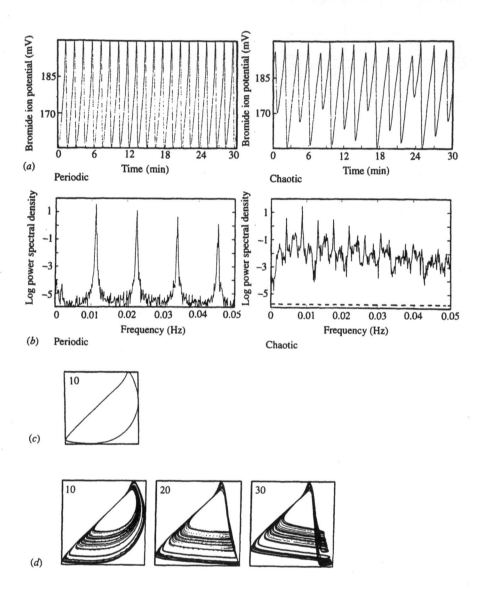

Figure B.4. Chemical chaos: the Belousov–Zhabotinsky chemical reaction. (*a*) Time series of the bromide ion potential. (*b*) Power spectra associated with the time series in (*a*). (*c*) Periodic limit-cycle attractor. (*d*) Strange attractor plots: reconstructed using various delays. (After Roux J-C, Simoyi R H and Swinney H L 1983 Observations of a strange attractor *Physica* D **8** 257–66. With kind permission of Elsevier Science—NL, Sara Burgerhartstraat 25, 1055 KV Amsterdam, The Netherlands.)

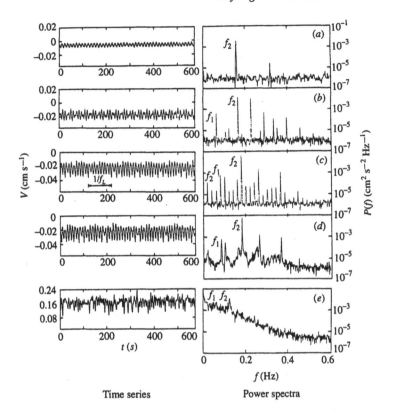

Figure B.5. Chaos in the Taylor–Couette experiment. (Figure 6.16 contains a schematic diagram of this system.) (After Gollub and Benson (1980). Reproduced with permission of Cambridge University Press.).

details of the experiment.) The figures are arranged from low to high Reynolds numbers (top to bottom), where the Reynolds numbers, R^*, have been further non-dimensionalized by dividing the flow Reynolds numbers by that Reynolds number at which broad-band chaos is observed. At low Reynolds numbers, one frequency is present at f_1 together with its associated harmonics at nf_1 (a); as the Reynolds number increases a second frequency emerges at f_2 (b); and a further increase brings about f_3 ((c), (d)). The onset of broad-band chaos is illustrated in figure B.5(e), where a sudden spectral broadening occurs as R^* passes through unity.

B.6 Rayleigh–Benard convection

Time series and power spectra obtained by Gollub and Benson (1980) for a Rayleigh–Benard convection experiment are plotted in figure B.6 (refer back to figure 6.6 for a schematic diagram of the phenomenon.) The flow Rayleigh number increases from top to bottom in the figure. Five cases are shown: periodic (a); quasi-periodic (two incommensurate frequencies (b); phase locked frequencies (c); non-periodic with distinct

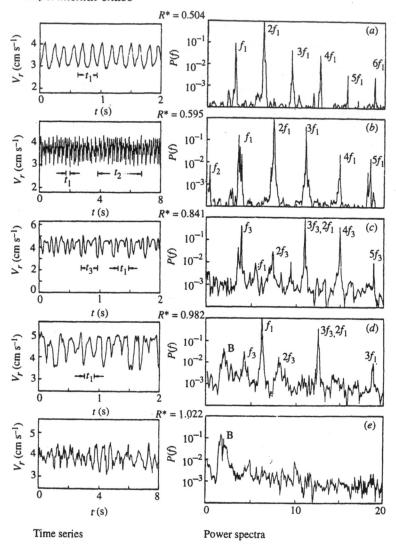

Figure B.6. Rayleigh–Benard convection. (Figure 6.6 contains a schematic diagram of this system.) (After Gollub and Swinney (1975).)

peaks (*d*); and strongly non-periodic with no sharp peaks within the broad-band noise of the spectral plot (*e*).

B.7 The ring cavity

Ikeda *et al* (1980) investigated the behaviour of transmitted light from a ring cavity which contained a nonlinear dielectric medium. They found stationary, periodic and chaotic states occurring as the intensity of the incident light was increased. Figure B.7(*a*) gives a schematic diagram of the ring cavity containing the nonlinear dielectric medium; and

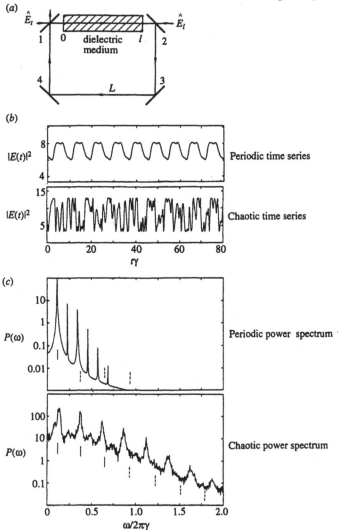

Figure B.7. Observed chaos in a ring cavity. (*a*) A sketch of a ring cavity containing a nonlinear dielectric medium. (*b*) Time series of a periodic and chaotic signal. (*c*) Power spectra of the periodic and chaotic signals given in (*b*). (After Ikeda *et al* (1980). Reproduced by kind permission of the authors.).

figure B.7(*b*) shows time series of the measurement of the light output from the cavity for two different settings of control parameters: one for a periodic case, and one for a chaotic case. The power spectra corresponding to the times series are given below each time series in figure B.7(*c*).

Appendix C

Solutions

C.1 Solutions to chapter 2

Q2.1 See the text.

Q2.2 Line: $D_E = 1$, $D_T = 1$ $N.\varepsilon = 1$ hence $1/\varepsilon = N$ and $D_S = \log(N)/\log(1/\varepsilon) = \log(1/\varepsilon)/\log(1/\varepsilon) = 1$
   Not fractal as $D_S = D_T$
   Area: $D_E = 2$, $D_T = 2$, $N\varepsilon^2 = 1$ hence $1/\varepsilon^2 = N$ and
   $D_S = \log(N)/\log(1/\varepsilon) = 2\log(1/\varepsilon)/\log(1/\varepsilon) = 2$
   Again not fractal as $D_S = D_T$.

Q2.3 (a) $D_E = 2$ and $D_T = 1$
   (b) $N = 5$, $\varepsilon = 1/3$. $D_S = \log(N)/\log(1/\varepsilon) = \log(5)/\log(3) = 1.4649\ldots$
   (c) $N = 3$, $\varepsilon = 1/2$. $D_S = \log(N)/\log(1/\varepsilon) = \log(3)/\log(2) = 1.5849\ldots$
   (d) Use figure 2.13 to give yourself some ideas. Find D_S in the same manner as (b) and (c).

Q2.4 Left-hand set: $N = 2$, $\varepsilon = 3/8$. $D_S = \log(N)/\log(1/\varepsilon) = \log(2)/\log(\frac{8}{3}) = 0.7066\ldots$
   Right-hand set: $N = 3$, $\varepsilon = 1/5$. $D_S = \log(N)/\log(1/\varepsilon) = \log(3)/\log(5) = 0.6826\ldots$

Q2.5 See figure C.1.
   $N = 5$, $\varepsilon = 1/3$. $D_S = \log(N)/\log(1/\varepsilon) = \log(5)/\log(3) = 1.4649\ldots$

Q2.6 (a) $N = 20$, $\varepsilon = 1/4$. $D_S = \log(N)/\log(1/\varepsilon) = \log(20)/\log(4) = 2.1609\ldots$
   (b) Here D_S is greater than 2 due to overlapping parts of the curve.
   (c) Use figure 2.15 for ideas.

Q2.7 $N = 8$, $\varepsilon = 1/3$. $D_S = \log(N)/\log(1/\varepsilon) = \log(8)/\log(3) = 1.8927\ldots$

Q2.8 $N = 4$, $\varepsilon = 1/4$. $D_S = \log(N)/\log(1/\varepsilon) = \log(4)/\log(4) = 1$.
   $D_T = 0$ and $D_E = 2$.
   It is a fractal as $D_S > D_T$

Q2.9 (a) See figure C.2.

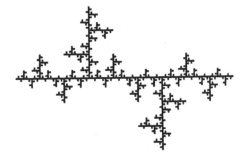

Figure C.1. See solution to Q2.5.

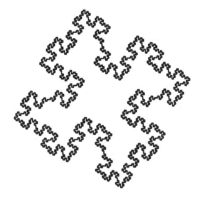

Figure C.2. See solution to Q2.9(*a*).

(*b*) The quadratic Koch island has unit area, as at each stage in the construction the amount of island area cut out is equal to that appended.

(*c*) a^2.

C.2   Solutions to chapter 3

Q3.1 See the text.

Q3.2 To cover the square lamina, shown in figure C.3, we could use one cube of side length 8 units; four cubes of side length 4 units; sixteen cubes of side length 2 units and so on.

Table C.1. See solution to Q3.2.

Cube side length (δ)	8	4	2	1	0.5
Number of cubes required to cover lamina (N)		1	4	16	64	256

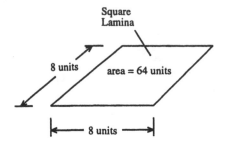

Figure C.3. The plane lamina of Q3.2.

Tabulating these results we get table C.1.
Plot $\log(N)$ against $\log(1/\delta)$,

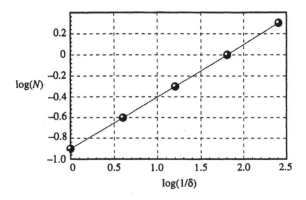

Figure C.4. See solution to Q3.2.

Plotting $\log(N)$ against $\log(1/\delta)$ as shown in figure C.4 (using any base, the plot above is base 10) we find that the slope may be found using any two of the ($\log(N)$, $\log(1/\delta)$) pairs, e.g.

$$D_B = \frac{\log(N_2) - \log(N_1)}{\log(1/\delta_2) - \log(1/\delta_1)} = \frac{\log(16) - \log(4)}{\log(\frac{1}{2}) - \log(\frac{1}{4})} = \frac{\log(4)}{\log(2)} = \frac{2\log(2)}{\log(2)} = 2.$$

Alternatively, we know $V^* = $ area $= 64$ units, hence we may obtain D_B using any set of N and δ, as follows:

$$D_B = \frac{\log(N) - \log(V^*)}{\log(1/\delta)} = \frac{\log(16) - \log(64)}{\log(\frac{1}{2})} = \frac{\log(\frac{1}{4})}{\log(\frac{1}{2})} = 2.$$

Q3.3 (*a*) Cover the Cantor set with boxes as shown in figure C.5.
   Tabulate the number of boxes required for each side length as in table C.2.
   As with question 3.2, plotting $\log(N)$ against $\log(1/\delta)$ (figure C.6), the slope of the resulting line is the box counting dimension, D_B. Taking two sets of N and δ, we obtain

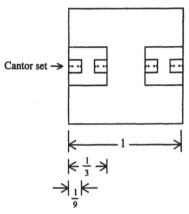

Cantor set →

Figure C.5. See solution to Q3.3(a).

Table C.2. See solution to Q3.3(a).

Cube side length (δ)	1	$\frac{1}{3}$	$\frac{1}{9}$	$\frac{1}{27}$	$\frac{1}{81}$
Number of cubes required to cover lamina (N)	1	2	4	8	16

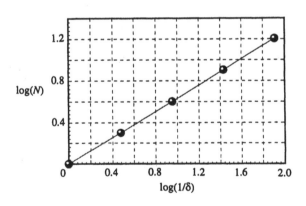

Figure C.6. See solution to Q3.3(a).

D_B as follows

$$D_B = \frac{\log(N_2) - \log(N_1)}{\log(1/\delta_2) - \log(1/\delta_1)} = \frac{\log(8) - \log(2)}{\log(\frac{1}{27}) - \log(\frac{1}{3})} = \frac{\log(4)}{\log(9)} = \frac{2\log(2)}{2\log(3)} = 0.6309\ldots.$$

(b) Generate a random Cantor set as shown at the top of figure 3.1. The number of boxes required to cover the set at each stage is the same as in part (a). Hence, again

$$D_B = 0.6309\ldots.$$

Table C.3. See solution to Q3.3(c).

Box side length (δ)	1	$\frac{2}{3}$	$\frac{1}{2}$	$\frac{1}{3}$	$\frac{1}{5}$	$\frac{1}{6}$	$\frac{1}{7}$	$\frac{1}{8}$	$\frac{1}{9}$	$\frac{1}{10}$	$\frac{1}{12}$	$\frac{1}{17}$	$\frac{1}{27}$
Number of boxes required (N)	1	2	2	2	4	4	4	4	4	8	8	8	8

(c) Attempt to cover the set with boxes of various side lengths. Some examples are tabulated as shown in table C.3.

Plotting $\log(N)$ against $\log(1/\delta)$ gives the graph below in figure C.7.

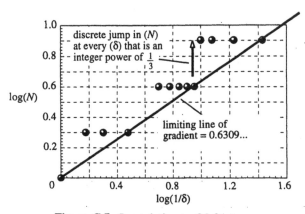

Figure C.7. See solution to Q3.3(c).

Notice that the number N required jumps discretely at each power of $\frac{1}{3}$. This highlights one of the idiosyncracies of the technique—*beware of this effect in practice*!

Q3.4 The large filled circles on the plot in figure C.8 are for step lengths of 1, $\frac{1}{3}$, $\frac{1}{9}$, $\frac{1}{27}$ and $\frac{1}{81}$, i.e. integer powers of one-third of the base length. The smaller triangles show various attempts at finding L using step sizes other than integer powers of $\frac{1}{3}$. As with question 3.3(c) the reader is warned to beware of this type of effect when using this technique in practice.

Q3.5 For the Spanish–Portuguese border

$$D_D = 1 - S = 1 - \frac{\log(L_2) - \log(L_1)}{\log(\lambda_2) - \log(\lambda_1)} = 1 - \frac{2.75 - 2.90}{2.5 - 1.5} = 1.15.$$

Similarly, for the west coast of Britain $D_D = 1.22$. (Note that your estimate of D_D may differ slightly due to your reading of the graph.) D_D for the circle tends to unity at small scales (i.e. small λ).

Q3.6 Counting the number of boxes that contain a piece of the coastline for each of the four box sizes gives the results shown in table C.4.

The base of the grid has been taken as unit length, hence box side lengths are the reciprocals of the number of boxes on the base. (Note that any measurement will do:

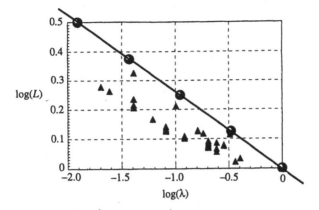

Figure C.8. See solution to Q3.4.

Table C.4. See solution to Q3.6.

Box side length (δ)	$\frac{1}{10}$	$\frac{1}{15}$	$\frac{1}{20}$	$\frac{1}{30}$
Number of boxes required to cover lamina (N)	42	70	98	163

for example you could directly measure the base in millimetres.) Plotting $\log(N)$ against $\log(1/\delta)$ gives figure C.9. The best-fit line is shown and its gradient gives the box counting dimension as $D_B = 1.25$. (Note that your answer will almost certainly differ slightly due to your own judgement in the box counting process—count all boxes which contain or touch the coastline.)

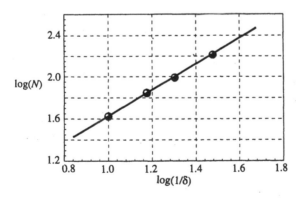

Figure C.9. See solution to Q3.6.

Q3.7 Refer to figure 3.12 in the text.

Q3.8 Begin at A, B and C in turn and use the inswing, outswing and alternate methods to produce nine sets of $\log(L)$ against $\log(\lambda)$. These are plotted below in figure C.10.

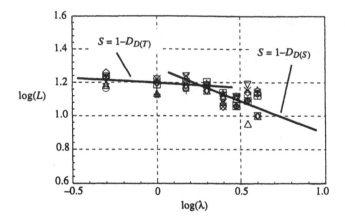

Figure C.10. See solution to Q3.8.

The plots of the best fit lines give the following estimates of the textural dimension and structural dimension: $D_{D(T)} = 1.06$ and $D_{D(S)} = 1.38$.

Notice that in this question you will have to decide which way to walk around the boundary (i.e. clockwise or anticlockwise), and for the alternate method whether to start with an inswing or outswing. You are left to investigate these effects for yourself. You are also urged to try alternative starting positions.

Q3.9 In the same manner as Q3.8, the nine sets of $\log(L)$ against $\log(\lambda)$ are plotted below for particle number 1. From the plot in figure C.11 the best fit lines through the points give $D_{D(T)} = 1.14$ and $D_{D(S)} = 1.38$.

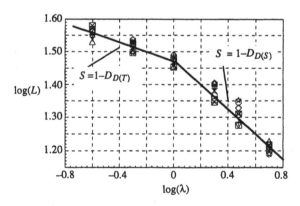

Figure C.11. See solution to Q3.9.

Repeating the procedure for particle number 2 gives similar dimension results.

Q3.10 Plot the measured length, L, of the coastline against the step length, λ. This is shown in figure C.12 with all the lengths normalized against the direct distance between Mallaig and Portpatrick. From the plot the slope S is -0.25, giving a divider dimension, D_D, of 1.25

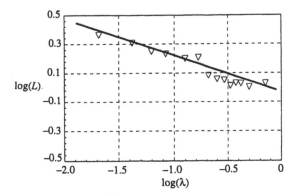

Figure C.12. See solution to Q3.10.

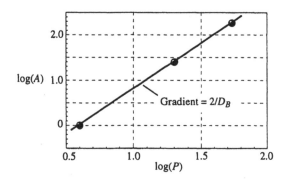

Figure C.13. See solution to Q3.12.

Q3.11 Repeat the procedures of Q3.6 and Q3.10. You can also try other boundary fractals.

Q3.12 (*a*) Choose any side lengths you wish. Here we have chosen $L = 1$, $L = 5$, and $L = 13.574$. Table C.5 lists the perimeters and areas of the squares.

Table C.5. See solution to Q3.12(*a*).

Square side length L	1	5	13.57
Perimeter (P)	4	20	54.28
Area (A)	1	25	184.14

Plotting $\log(A)$ against $\log(P)$ we get the graph shown in figure C.13.

The gradient is 2 ($= 2/D_B$). Hence, $D_B = 1$, as we would expect since the boundary is made up of straight lines.

(*b*) Repeat the procedure for part (*a*); this time the perimeter is $3L$ and the area is $\left(\sqrt{3}/2\right) L^2$, where L is the side length of the triangle. As with part (*a*) you should find that $D_B = 1$.

(c) This is a bit trickier. If the initiator length is a, then the hypervolume of each of the three constituent Koch curves on the boundary is

$$V^* = a^{D_B} = N\delta^{D_B}.$$

Now consider the iterative generation of the Koch curve coastline. The coastline length at the kth iteration is three times that for a single Koch curve, i.e.

$$P_k = 3\,(N\delta)$$

where N is the number of segments of length δ required to make up the kth iteration of the Koch prefractal. We may also think of δ as the side length of measuring boxes placed on the fractal Koch curve, i.e. $\delta = \left(\frac{1}{3}\right)^k a$. The measured length of the perimeter using δ is then

$$P_\delta = 3\left[\left(\frac{a}{\delta}\right)^{D_B}\delta\right] = 3\left[(a)^{D_B}(\delta)^{1-D_B}\right].$$

Now if we choose δ small enough then the measured area will converge to

$$A = \frac{2}{5}\sqrt{3}a^2$$

thus the perimeter–area ratio is

$$R = \frac{(P)^{1/D_B}}{\sqrt{A}} = \left[3\left[\left(\frac{a}{\delta}\right)^{D_B}\delta\right]\right]^{1/D_B}\left(\tfrac{2}{5}\sqrt{3}a^2\right)^{-1/2} = C_1(\delta)^{(1/D_B)-1}.$$

We see that the a terms in the above expression cancel out, leaving the right-hand term, where C_1 is a constant. Thus, for a specific δ (as long as δ is small enough to measure A accurately), the area–perimeter ratio, R, is a constant regardless of the size of the Koch snowflake. Or to put it another way, the perimeter and area obey the following scaling relationship:

$$(P)^{2/D_B} = C_2 A$$

where C_2 is a constant. Hence, a logarithmic perimeter–area plot of A against P results in a line of slope $2/D_B$. We can then obtain D_B from the plot and we already know from chapter 2 that $D_B = D_S = 1.2618\ldots$.

You can try verifying this result by measuring a few Koch snowflakes of different sizes drawn on a fine grid—if you have the patience!

(d) You can try detailed maps of actual island groups, photographs of clouds, detailed contour plots of rough terrain, etc. Draw a fine grid over the group of island fractals and measure the perimeter P (= total number of boxes containing part of the boundary multiplied by the grid box length, δ) and area A (= total number of boxes containing part of the island multiplied by δ^2). Plot the logarithm of A against P for each island fractal and find the resulting slope of the best fit line through the points. Hence find D_B.

Q3.13 Try to think of natural objects which reveal more of the 'same kind of structure' as one zooms in. See the chapter summary and further reading section at the end of the chapter for further examples.

C.3   Solutions to chapter 4

Q4.1 See the text.

Q4.2 Dimension of the trace function $= 2 - H = 2 - 0.4 = 1.6$. Dimension of the zero set $= 1 - H = 1 - 0.4 = 0.6$.

Q4.3 and Q4.4 Use the first two columns of table Q4.3 for the x co-ordinate trace and columns three and four for the y co-ordinate time trace. The x co-ordinate time trace and the x–y trajectory co-ordinates are given in table C.6 and figures C.14 and C.15.

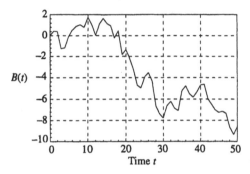

Brownian motion trace using the first two columns
of random numbers

Figure C.14. See solution to Q4.3.

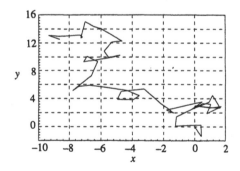

Brownian motion trajectory in the plane

Figure C.15. See solution to Q4.4.

Q4.5 $D = N + 1 - H$, hence $2.32 = 2 + 1 - H$, giving $H = 0.68$.

Q4.6 Divide the plot up into smaller boxes: two examples are shown in figure C.16. Count the number of filled boxes, N, corresponding to each box side length, δ. Plot $\log(N)$ against $\log(\delta)$, and estimate the gradient of the best-fit line to find the fractal dimension of the curve (see table C.7).

Table C.6. See solution to Q4.3 and Q4.4.

Time (arbitrary units)	Brownian motion trace 1 (x co-ordinate)	Brownian motion trace 2 (y co-ordinate)
0	0	0
1	0.430 68	−1.376 922
2	0.404 495	0.263 957
3	−1.245 137	0.127 249
4	−1.169 77	1.433 329
5	−0.120 138	2.556 453
6	0.524 479	3.324 174
7	0.896 992	3.263 974
8	1.048 129	3.190 999
9	0.789 104	1.896 113
10	1.740 125	2.884 574
11	1.016 953	2.689 739
12	0.029 144	3.014 895
13	1.137 632	2.725 797
14	1.657 231	2.680 577
15	1.168 763	4.466 291
16	0.946 334	3.519 221
17	−0.220 022	2.449 372
18	0.477 012	3.564 136
19	−1.825 248	2.371 557
20	−1.336 625	1.964 285
21	−2.168 502	3.270 266
22	−3.235 152	5.338 224
23	−4.710 029	4.998 106
24	−4.966 364	3.881 967
25	−3.924 953	3.843 977
26	−3.548 751	4.389 035
27	−4.338 681	5.151 283
28	−6.661 315	5.879 183
29	−7.379 107	5.715 638
30	−7.780 865	5.093 572
31	−6.596 326	7.238 712
32	−6.212 215	9.364 244
33	−6.860 441	10.091 02
34	−7.078 316	9.291 66
35	−5.210 226	10.002 692
36	−4.766 208	10.296 409
37	−5.518 595	9.936 981
38	−5.816 649	10.843 783
39	−5.329 628	12.112 495
40	−4.694 242	12.321 939
41	−4.633 346	12.273 093
42	−5.976 301	14.012 943
43	−6.478 266	14.394 799
44	−6.992 421	15.039 488
45	−7.246 672	12.718 57
46	−7.139 64	12.860 104
47	−7.312 652	13.148 318
48	−8.636 134	12.953 566
49	−9.353 725	13.105 252
50	−8.610 945	12.514 164

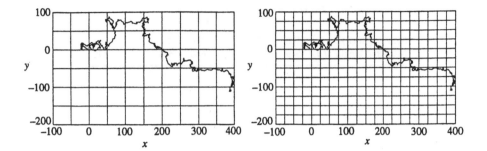

Figure C.16. See solution to Q4.6.

Table C.7. See solution to Q4.6.

Box side length (δ)	100	50	25	12.5
Number of boxes (N) required to cover trajectory	8	18	38	93

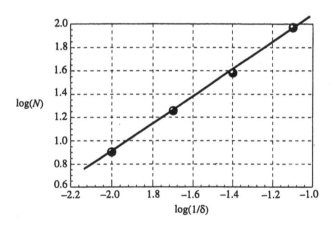

Figure C.17. See solution to Q4.6.

Plotting $\log(N)$ against $\log(1/\delta)$ gives the graph shown in figure C.17.

From the slope of the best-fit line the box counting dimension is $D_B = 1.18$. The value of D_B you obtain will vary slightly according to the box size you choose and the accuracy of your counting. Note that the trajectory in figure Q4.6 is a simulated fBm trajectory with $H = 0.8$, hence we would expect to find $D_B = 1/H = 1.25$.

Q4.7 Walk along the fBm trajectory of figure Q4.5 and find the measured length for various step lengths. Table C.8 gives some examples. Then plot $\log(L)$ against $\log(\lambda)$ and find the slope, S, of the best-fit line connecting the plotted points, $D_D = 1 - S$. These results give D_D to be approximately 1.22 (see figure C.18). (As with Q4.6, you should obtain an answer in the region of 1.25 for D_D.)

Table C.8. See solution to Q4.7.

Step length (λ)	100	67	50	36	27	18	9
Measured length (L)	549	567	576	636	681	772	872

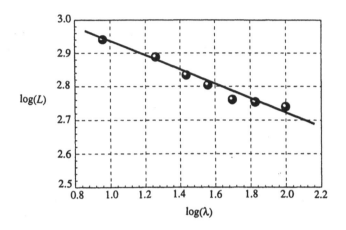

Figure C.18. See solution to Q4.7.

Notice that in this question the fractal curve crosses itself, unlike the coastline curves of chapter 3. When walking along the curve, use your own judgement to end each step at the nearest point (in time) along the curve.

Q4.8 Compare graphical output with figures 4.1 and 4.2. Remember you may use a non-Gaussian probability distribution to approximate Brownian motion (see figure 4.5).

Q4.9 (a) Produce a very long time series. Moving along the time series compare each point on it, $B(t)$, to $B(t + T_s)$. Calculate the absolute separation for each pair, $|\Delta B| = B(t + T_s) - B(t)$, and find the average value over the whole time series. Check the relationship $\overline{|\Delta B|} \propto T_s^{1/2}$ by plotting $\left[\overline{|\Delta B|}\right]^2$ against T_s for various value of T_s.

(b) Similar to part (a), calculate the standard deviation of the Brownian particle cloud, σ_c, at each time step. Verify the relationship $\sigma_c \propto t^{1/2}$ by plotting $[\sigma_c]^2$ against time (see figure 4.4(c)). From the plot you should be able to determine a diffusion coefficient from the slope.

Q4.10 Compare graphical output with figure 4.12.

Q4.11 (a) A similar method to Q4.9(a). This time check the relationship $\overline{|\Delta B|} \propto T_s^H$ by plotting $\left[\overline{|\Delta B|}\right]^{1/H}$ against T_s. The accuracy of your results in this question and part (b) will depend upon the size of memory you use in the computation of the fBm.

(b) Again similar to Q4.9(b). Verify the relationship $\sigma_c \propto t^H$ by plotting $[\sigma_c]^{1/H}$ against time.

Q4.12 (a) Alter your program to locate the crossing points of the Brownian trace on the time axis. You will first have to have to check for crossing points, knowing that

Table C.9. See solution to Q5.2.

Iteration Number	Control parameter setting				
n	$a = 0.9$	$a = 2.8$	$a = 3.18$	$a = 3.5$	$a = 4$
0	0.2	0.2	0.2	0.2	0.2
1	0.144	0.448	0.5088	0.56	0.64
2	0.1109...	0.6924...	0.7947...	0.8624	0.9216
3	0.0887...	0.5963...	0.5187...	0.4153...	0.2890...
4	0.0727...	0.6740...	0.7938...	0.8499...	0.8219...
5	0.0607...	0.6152...	0.5203...	0.4464...	0.5854...
6	0.0513...	0.6628...	0.7936...	0.8649...	0.9708...
7	0.0438...	0.6257...	0.5207...	0.4087...	0.1133...
8	0.0377...	0.6557...	0.7936...	0.8458...	0.4019...
9	0.0326...	0.6321...	0.5208...	0.4562...	0.9615...
10	0.0284...	0.6511...	0.7936...	0.8683...	0.1478...
11	0.0248...	0.6360...	0.5208...	0.4002...	0.5039...
12	0.0218...	0.6481...	0.7936...	0.8401...	0.9999...
13	0.0192...	0.6385...	0.5208...	0.4700...	0.0002...
14	0.0169...	0.6462...	0.7936...	0.8718...	0.0009...
15	0.0150...	0.6400...	0.5208...	0.3910...	0.0039...
16	0.0133...	0.6450...	0.7936...	0.8334...	0.0156...
17	0.0118...	0.6410...	0.5208...	0.4858...	0.0617...
18	0.0105...	0.6442...	0.7936...	0.8743...	0.2317...
19	0.0093...	0.6417...	0.5208...	0.3846...	0.7121...
20	0.0083...	0.6437...	0.7936...	0.8284...	0.8200...
Periodicity:	period 1 (decay to origin)	tending to period 1	period 2	tending to period 4 (eventually !)	aperiodic = chaos

crossing involves consecutive points on the trace changing sign, i.e. $+$ to $-$, or vice versa. Interpolate between consecutive points with changed sign to approximate the crossing points.

(b) Use the distance between consecutive crossing points in part (a) as the steps in the Levy flight. Randomly select each step angle between 0 and 2π rad.

Q4.13 Follow the method given in section 4.6 and figure 4.16(a). Compare your graphical output with figure 4.16(b).

Q4.14 $\phi = 2.6$ and $D = [5/2 - \phi/2]$, giving $D = 1.2$ and $\phi = 1 + 2H$ giving $H = 0.8$.

C.4   Solutions to chapter 5

Q5.1 See the text.

Q5.2 See table C.9.

Q5.3 See figure C.19.

Q5.4 See figure C.20.

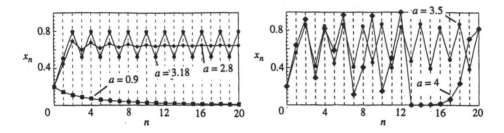

Figure C.19. See solution to Q5.3.

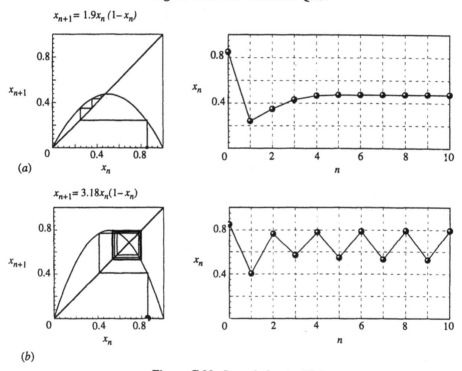

$x_{n+1} = 1.9x_n(1-x_n)$

(a)

$x_{n+1} = 3.18x_n(1-x_n)$

(b)

Figure C.20. See solution to Q5.4.

Q5.5 The post transient solutions for $a = 0.9$, 2.8, 3.18 and 3.5 are shown as circles on the bifurcation diagram of figure C.21. The $a = 4$ chaotic sequence of solutions is spread over the interval $0 < x < 1$. This is shown by a thick black line in the diagram.

Q5.6

$$\frac{a_2 - a_1}{a_3 - a_2} = \frac{3.449\,490 - 3.0}{3.544\,090 - 3.449\,490} = 4.751\,479\ldots$$

$$\frac{a_3 - a_2}{a_4 - a_3} = \frac{3.544\,090 - 3.449\,490}{3.564\,407 - 3.544\,090} = 4.656\,199\ldots$$

$$\frac{a_4 - a_3}{a_5 - a_4} = \frac{3.564\,407 - 3.544\,090}{3.568\,759 - 3.564\,407} = 4.668\,428\ldots$$

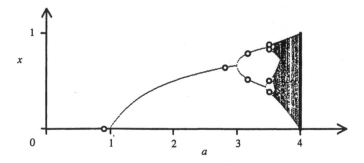

Figure C.21. See solution to Q5.5.

$$\frac{a_5 - a_4}{a_6 - a_5} = \frac{3.568\,759 - 3.564\,407}{3.569\,692 - 3.568\,759} = 4.664\,523\ldots$$

$$\frac{a_6 - a_5}{a_7 - a_6} = \frac{3.569\,692 - 3.568\,759}{3.569\,891 - 3.569\,692} = 4.688\,442\ldots.$$

Notice that the ratios fluctuate around the Feigenbaum number of $4.669\,201\ldots$; only as k tends to infinity do we realize δ.

Q5.7 See table C.10. The two $a = 2.8$ iterations lead to the fixed point attractor at $0.6428\ldots$, thus both initial conditions lie in the basin of attraction for this attractor. However, the initial condition of $x_0 = 0.5$, for the $a = 4$ case, throws the orbit directly onto the repelling fixed point at the origin, whereas the $x_0 = 0.8$ initial condition leads to a chaotic sequence of iterates.

If we iterate backwards, from the initial condition, setting $x_{n+1} = 0.5$, then solving $x_{n+1} = 4x_n(1 - x_n)$ we obtain two solutions for x_n, namely

$$\frac{1 + \sqrt{\frac{1}{2}}}{2} \quad \text{and} \quad \frac{1 - \sqrt{\frac{1}{2}}}{2}.$$

We may repeat this back-iteration for these two points to obtain four points and so on; each time we obtain twice as many x values which lead to a steady state orbit at $x = 0$. Thus, after infinite back iterations we have an infinite number of x values which lead to a steady state orbit at $x = 0$. These are intermixed with an infinite number of x values which lead to a chaotic orbit for $a = 4$, such as the one we observed for the $x = 0.8$ initial condition.

Q5.8 See table C.11.

Q5.9 Apart from the first few transient iterates, the Henon map iterates should all appear to fall somewhere on the attractor plotted in figure 5.6(*b*). A periodic attractor for $a = 0.9$, $b = 0.3$ is given in figure 5.6(*a*), others may be found for low values of b.

Q5.10 See table C.12.

Q5.11 Julia set.

Q5.12 See table C.13 on p 222.

Table C.10. See solution to Q5.7.

Iteration number	Control parameter setting			
n	$a = 2.8$		$a = 4$	
0	0.5	0.8	0.5	0.8
1	0.7	0.448	1	0.64
2	0.588	0.6924...	0	0.9216
3	0.6783...	0.5963...	0	0.2890...
4	0.6109...	0.6740...	0	0.8219...
5	0.6655...	0.6152...	0	0.5854...
6	0.6232...	0.6628...	0	0.9708...
7	0.6574...	0.6257...	0	0.1133...
8	0.6305...	0.6557...	0	0.4019...
9	0.6522...	0.6321...	0	0.9615...
10	0.6350...	0.6511...	0	0.1478...
11	0.6488...	0.6360...	0	0.5039...
12	0.6379...	0.6481...	0	0.9999...
13	0.6467...	0.6385...	0	0.0002...
14	0.6397...	0.6462...	0	0.0009...
15	0.6453...	0.6400...	0	0.0039...
16	0.6408...	0.6450...	0	0.0156...
17	0.6444...	0.6410...	0	0.0617...
18	0.6415...	0.6442...	0	0.2317...
19	0.6438...	0.6417...	0	0.7121...
20	0.6420...	0.6437...	0	0.8200...

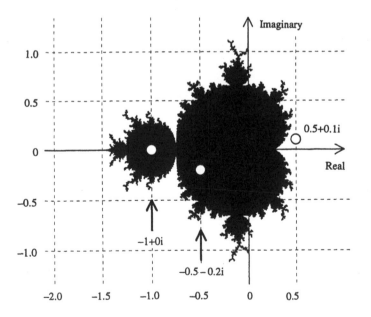

Figure C.22. See solution to Q5.13.

Table C.11. See solution to Q5.8.

Iteration number n	Control parameter setting					
	$a = 1.8$			$a = 2.1$		
	0.5	0.75	0.9	0.5	0.75	0.9
0	0.5	0.75	0.9	0.5	0.75	0.9
1	0.45	0.7159...	0.4505...	0.525	0.8352...	0.5256...
2	0.3600...	0.7183...	0.3609...	0.5770...	0.7248...	0.5783...
3	0.2110...	0.7187...	0.2125...	0.6788...	0.8393...	0.6812...
4	0.0493...	0.7188...	0.0503...	0.8189...	0.7154...	0.8208...
5	0.0006...	0.7188...	0.0007...	0.7585...	0.8379...	0.7550...
6	0.0000...	0.7188...	0.0000...	0.8311...	0.7186...	0.8330...
7	0.0000...	0.7188...	0.0000...	0.7339...	0.8385...	0.7298...
8	0.0000...	0.7188...	0.0000...	0.8392...	0.7171...	0.8394...
9		0.7188...		0.7157...	0.8383...	0.7151...
10		0.7188...		0.8380...	0.7178...	0.8378...
11		0.7188...		0.7185...	0.8384...	0.7188...
12		0.7188...		0.8385...	0.7175...	0.8386...
13		0.7188...		0.7172...	0.8383...	0.7170...
14		0.7188...		0.8383....	0.7176...	0.8383...
15		0.7188...		0.7177...	0.8384...	0.7178...
16		0.7188...		0.8384...	0.7175...	0.8384...
17		0.7188...		0.7175...	0.8384...	0.7175...
18		0.7188...		0.8383...	0.7176...	0.8383...
19		0.7188...		0.7176...	0.8384...	0.7176...
20		0.7188...		0.8384...	0.7176...	0.8384...
Periodicity:	period 1 (decay to zero)	period 1	period 1 (decay to zero)	period 2	period 2	period 2

Table C.12. See solution to Q5.10.

	$z_0 = -1 + 0i$		$z_0 = -1.5 + 1.5i$	
n	Real z	Imag. z	Real z	Imag. z
0	-1	0	1.5	1.5
1	0	0	-1	4.5
2	-1	0	-20.25	-9
3	0	0	328.06	364.50
4	-1	0	$-25\,236.24$	239 157.56
5	0	0	-56.559×10^9	-12.070×10^9
Behaviour	period 2		escape to infinity	

$z_0 = -1 + 0i$ is a member of the prisoner set.

Table C.13. See solution to Q5.12.

n	$c = -1 + 0i$		$c = 0.5 + 0.1i$		$c = 2 - 0.4i$		$c = -0.5 - 0.2i$	
	Real z	Imag. z	Real z	Imag. z	Real z	Imag. z	Real z	Imag. z
0	0	0	0	0	0	0	0	0
1	-1	0	0.5	0.1	2	-0.4	-0.5	-0.2
2	0	0	0.74	0.2	5.84	-2	-0.29	0
3	-1	0	1.0076	0.396	32.10	-23.76	-0.4159	-0.2
4	0	0	1.3584...	0.8980...	468.23	-1526.05	-0.367	-0.03364
5	-1	0	1.5389...	2.5398...	-2.11×10^6	-1.43×10^6	-0.3664...	-0.1753...
6	0	0	-3.5823...	7.9171...	2.41×10^{12}	6.03×10^{12}	-0.3964...	-0.0715...
7	-1	0	-49.3482...	-56.6243...			-0.3479...	-0.1432...
8	0	0	-3.06×10^6	-8.61×10^6			-0.3994...	-0.1002...
9	-1	0	864×10^{12}	527×10^{12}			-0.3504...	-0.1198...
10	0	0					-0.3915...	-0.1159...
11	-1	0					-0.3601...	-0.1091...
12	0	0					-0.3822...	-0.1213...
13	-1	0					-0.3686...	-0.1072...
14	0	0					-0.3756...	-0.1209...
15	-1	0					-0.3735...	-0.1091...
16	0	0					-0.372...	-0.1184...
17	-1	0					-0.3753...	-0.1117...
18	0	0					-0.3715...	-0.1160...
19	-1	0					-0.3753...	-0.1137...
20	0	0					-0.3720...	-0.1146...
Behaviour	period 2		escape to infinity		escape to infinity		period 1	
Julia set type	connected		disconnected		disconnected		connected	

Table C.14. See solution to Q5.15.

Large n	$c = 0 + 0i$		$c = -0.5 + 0i$		$c = -1.0 + 0i$		$c = -1.35 + 0i$		$c = -1.37 + 0i$	
	Real z	Imag. z	Real z	Imag. z	Real z	Imag. z	Real z	Imag. z	Real z	Imag. z
n	0	0	-0.3660...	0	0	0	-1.3464...	0	-1.1477...	0
$n+1$	0	0	-0.3660...	0	-1	0	0.4628...	0	-0.0525...	0
$n+2$	0	0	-0.3660...	0	0	0	-1.1358...	0	-1.3672...	0
$n+3$	0	0	-0.3660...	0	-1	0	-0.0599...	0	0.4993...	0
$n+4$	0	0	-0.3660...	0	0	0	-1.3464...		-1.1206...	0
$n+5$	0	0	-0.3660...	0	-1	0	0.4628...		-0.1140...	0
$n+6$	0	0	-0.3660...	0	0	0	-1.1358...		-1.3569...	0
$n+7$	0	0	-0.3660...	0	-1	0	-0.0599...		0.4714...	0
$n+8$	0	0	-0.3660...	0	0	0	-1.3464...	0	-1.1477...	0
Behaviour	decay to zero		period 1		period 2		period 4		period 8	

Q5.13 The points are circled on the plot in figure C.22.

Those points within the Mandelbrot set correspond to connected Julia sets, and those points outside correspond to disconnected Julia sets.

Q5.14 The Mandelbrot set is symmetric about the real axis, therefore, if $c = a + bi$ is outside the Mandelbrot set then so is $c = a - bi$. Thus, since we saw in Q5.12 and Q5.13 that $z_{n+1} = z_n^2 + 0.5 + 0.1i$ and $z_{n+1} = z_n^2 + 2 - 0.4i$ produce disconnected sets (i.e. lie outside the Mandelbrot set), then so must $z_{n+1} = z_n^2 + 0.5 - 0.1i$ and $z_{n+1} = z_n^2 + 2 + 0.4i$.

Q5.15 The post-transient orbits are given in table C.14 on p 223. For the period 8 case a few hundred iterations are required for the orbit to settle down onto the attractor. The map is following a period doubling bifurcation route to chaos as the control parameter, c, is increased along the negative real axis. A bifurcation diagram may be constructed by plotting the post-transient, real part of z against many values of c (keeping c real).

C.5 Solutions to chapter 6

Q6.1 See the text.

Q6.2 $\sum Forces = mass \times acceleration$: $F - F_r - F_s = ma$ where $F_r = rv$ and $F_s = sx^3$ and the forcing, F, takes the form $A_f \cos \omega t$. The equation of motion is then:

$$m\ddot{x} + r\dot{x} + sx^3 = A_f \cos \omega t.$$

Q6.3 See sections 6.2 and 6.3.

Q6.4 (*a*)The frequency, f, is halving, therefore the period T ($= 1/f$) is doubling. The system is, therefore, taking the period doubling route to chaos.

(*b*) Reulle–Takens, period doubling and intermittent routes.

Q6.5 You will need a FORTRAN compiler. The output file GEN.DAT lists four columns: t_i, x_i, y_i, z_i. It is probably simplest to import this data file into a graphics spreadsheet package and manipulate the data from there. Compare your results to figure 6.8.

Q6.6 As with Q6.5. This time explore the range $15 \leq r \leq 300$. See if you can find reverse period doubling sequences.

Q6.7 The modification to the GENERAL program required for the Rössler system is given in appendix 1. Explore the range $2 \leq c \leq 5.7$. See if you can find the bifurcation points on the period doubling sequence illustrated in figure 6.11 in the main text.

Q6.8 This is more involved and will require an increase in the number of ODEs specified as the parameter N in the program together with the insertion of additional lines of code for the extra ODE arrays. Compare your solutions with figures 6.12 and 6.15.

Q6.9 Use the first-order ODE form of the Duffing equation:

$$\dot{x} = y$$
$$\dot{y} = A_f \cos(t) - r\dot{x} - x^3.$$

Use this form to construct an EQNS subroutine for the Duffing equation. The best way to do this is to modify the Lorenz or Rössler routines. Note that the time variable (T) is already included in these routines (although unused). Use a time step of 0.05 and initial conditions given in the text. Compare your output with figures 6.2, 6.3 and 6.5.

Q6.10 Modify the program of Q6.9. Compare output with figure 6.14.

C.6 Solutions to chapter 7

Q7.1 See the text.

Q7.2 See section 7.2.

Q7.3 See section 7.3.

Q7.4 (*a*) Embedding dimension $m = 2(3) + 1 = 7$.
    (*b*) Whitney's embedding theorem.
    (*c*) $m > 2D_B \; (= 2 \times 2.12)$. Thus minimum m is 5.
    (*d*) $m > D_C \; (= 2.12)$. Thus minimum m is 3.
    (*e*) Use the correlation plot, increasing the embedding dimension until the $\log(r)$–$\log(C_r)$ curves reach a limiting slope. Find D_C, hence $m > D_C$.

Q7.5 (*a*) The $(2.813, 0, -9.543, -23.241)$ attractor is strange as it has a positive Lyapunov exponent of 2.813.
    (*b*) This exponent measures divergence/convergence along the trajectory itself, which is zero.
    (*c*) $(+, 0, -, -)$ and $(0, -, -, -)$

Q7.6
    (*a*) $\varepsilon_0 = 0.0032$; $\varepsilon_t = 0.0057$; $t = 0.005$.

$$\lambda = \frac{1}{t} \log_2 \left[\varepsilon_t / \varepsilon_0 \right] = 166.58 \text{ bits s}^{-1}.$$

(Note: you may not be able to directly calculate base two logarithms on your calculator. Convert from base ten logarithms as follows:

$$\log_2 [x] = \frac{\log_{10}[x]}{\log_{10}[2]}$$

or similarly from natural logarithms replacing $\log_{10}$ by ln in the above expression.)
    (*b*) λ is positive, therefore the system is chaotic.

Q7.7 (*a*) From Q7.6 we know that 166.58 bits of information is lost every second. One bit of information is lost every 0.006 00 s ($= 1/166.58$). Therefore, eight bits of information is lost in 0.0480 seconds. The orbital frequency is 500 Hz, hence one orbit takes 0.002 seconds. The number of orbits to lose eight bits of information is then $(0.0480/0.002) = 24.01$ orbits.
    Thus, current predictions would become invalid in 24.01 orbits on the attractor in a time of 0.0480 s.

Q7.8 Use equation (7.5) with base 2 logs.

ε_e is $2 \times 0.000\,05 = 0.0001$ and ε_a is 0.1304.

$$t = \frac{1}{\lambda} \log_2 \left[\varepsilon_a/\varepsilon_e \right] = \frac{1}{166.58} \log_2 \left[\frac{0.1304}{0.0001} \right] = 0.062 \text{ s or } 31.06 \text{ orbits.}$$

Q7.9 (a) The Lyapunov spectrum for the chaotic attractor is (2.813, 0, −9.543, −23.241).

First we require j, the largest integer which gives, $\sum_{i=1}^{j} \lambda_i > 0$. Increasing j in the summation we obtain

$$\sum_{i=1}^{1} \lambda_i = 2.813 \qquad \sum_{i=1}^{2} \lambda_i = 2.813 + 0 = 2.813$$

$$\sum_{i=1}^{3} \lambda_i = 2.813 + 0 + -9.543 = -6.730.$$

Thus $j = 2$, and $D_L = 2 + \frac{\sum_{i=1}^{2} \lambda_i}{|\lambda_3|} = 2 + \frac{(2.813+0)}{9.543} = 2.294$.

Q7.10 From the figure the limiting gradient, and hence D_C, is approximately 2.10.

Q7.11 (a) (0.0, 4.5); (1.1, 6.4); (2.3, 6.2); (3.3, 5.8); etc.

(b) Plot points in (a) (see figure C.23).

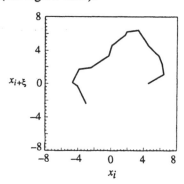

Figure C.23. See solution to Q7.11(b).

(c) (0.0, 3.3, 6.2); (1.1, 4.5, 5.8); (2.3, 6.4, 4.6); (3.3, 6.2, 3.3); (4.5, 5.8, 1.9).

Q7.12 (a) $D_{C(max)} = \frac{2\log(10\,000)}{\log(1/0.1)} = 8$.

(b) Filtering, data sampling rate, noise, etc.

Q7.13 See the discussion of correlation dimension in section 7.5. Write a routine based on equation 7.8(a) to calculate the correlation integral, for various values of hypersphere radius. Essentially you want to visit each of the data points in turn and count how many other points on the attractor are within a radius r. At each visited point, count the number of points contained within the hypersphere over a range of r. Calculate C_r for all the radii at the end of the time series. Then plot $\log(C_r)$ against $\log(r)$ and use the slopes to estimate D_C. Compare your answer to figure 7.4.

Q7.14 Use the x–time series output contained in the second column of the output data file GEN.DAT to generate the phase portraits. Plot the time series x_i against its delayed self $x_{i+\xi}$. You should be able to do this with relative ease on most basic spreadsheet packages with graphic capabilities. Compare your figures with figure 7.9(b).

Q7.15 See section 7.6 and figure 7.8 for the mutual information function. Try it for a 2D reconstruction first of all (equation (7.20)). Compare your answer to figure 7.9(a). If you feel comfortable write a routine to calculate the mutual information for higher dimensional reconstructions.

Q7.16 This follows on from Q7.13–15. This time remember we use the reconstructed variables x_i, $x_{i+\xi}$, $x_{i+2\xi}$ etc instead of x_i, y_i and z_i as we did in Q7.13. Compare your results with figure 7.10.

Q7.17 Modify the general program for Rössler output as outlined in appendix 1.

Q7.18 Compare your answers to figure 7.12.

References

The chapter(s) in which the reference is cited is given in parenthesis at the end of each reference.

Abraham N B, Albano A M, Das B, De Guzman G, Yong S, Gioggia R S, Puccioni G P and Tredicce J R 1986 Calculating the dimension of attractors from small data sets *Phys. Lett.* **114A** 217–21 (Ch 7)

Abraham N B, Arecchi F T and Lugiato L A (eds) 1988 *Instabilities and Chaos in Quantum Optics II* (New York: Plenum) (Ch 6)

Abraham R H and Shaw C D 1982 *Dynamics—The Geometry of Behaviour. Part 1: Periodic Behaviour* (Santa Cruz, CA: Aerial) (Ch 6)

——1983 *Dynamics—The Geometry of Behaviour. Part 2: Chaotic Behaviour* (Santa Cruz, CA: Aerial) (Ch 6)

——1984 *Dynamics—The Geometry of Behaviour. Part 3: Global Behaviour* (Santa Cruz, CA: Aerial) (Ch 6)

——1988 *Dynamics—The Geometry of Behaviour. Part 4: Bifurcation Behaviour* (Santa Cruz, CA: Aerial) (Ch 6)

Acuna J A, Ershaghi I and Yortsos Y C 1995 Practical application of fractal pressure-transient analysis in naturally fractured reservoirs *SPE Formation Evaluation* 173–9 (Ch 3)

Acuna J A and Yortsos Y C 1995 Application of fractal geometry to the study of networks of fractures and their pressure transient *Water Resources Res.* **10** 527–40 (Ch 3)

Addison P S 1995 On the characterization of non-linear oscillator systems in chaotic mode *J. Sound Vibration* **179** 385–98 (Ch 6)

——1996 A method for modelling dispersion dynamics in coastal waters using fractional Brownian motion *IAHR J. Hydraul. Res.* **34** 549–61 (Ch 4)

Addison P S and Bo Q 1996 Modeling fractal diffusion on the ocean surface *Flow Modeling and Turbulence Measurements VI* ed C-J Chen, S Shih, J Lienau and R J Kung *Proc. 6th Int. Symp. Flow Modeling and Turbulence Measurements (Tallahassee, FL, USA, 1996)* (Rotterdam: Balkema) pp 703–9 (Ch 4)

Addison P S, Chan A H C, Ervine D A and Williams K J 1992 Observations on numerical method dependent solutions of a modified Duffing oscillator *Commun. Appl. Numer. Methods* **8** 519–28 (Ch 6)

Addison P S and Low D J 1996 Order and chaos in the dynamics of vehicle platoons *Traffic Eng. Control* **37** 456–9 (Ch 6)

Adiletta G, Guido A R and Rossi C 1996a Chaotic motions of a rigid rotor in short journal bearings *Nonlinear Dynam.* **10** 251–69 (Ch 6)

——1996b Non-periodic motions of a Jeffcott rotor with non-linear elastic restoring forces *Nonlinear Dynam.* **11** 37–59 (Ch 6)

Aharony A and Feder J (eds) 1990 *Fractals in Physics: Essays in Honour of Benoit B Mandelbrot* (Amsterdam: North-Holland) (Ch 3)

Albano A M, Mees A I, De Guzman G C and Rapp P E 1987 Data requirements for reliable estimation of correlation dimensions *Chaos in Biological Systems* ed H Degn, A V Holden and L F Olsen (*NATO ASI Series A: Life Sciences vol 138*) (New York: Plenum) pp 207–20 (Ch 7)

Albano A M, Muench J, Schwartz C, Mees A I and Rapp P E 1988 Singular-value decomposition and the Grassberger–Procaccia algorithm *Phys. Rev.* A **38** 3017–26 (Ch 7)

Ali M, Gennert M A and Clarkson T G 1992 Analysis, generation and compression of pavement distress images using fractals *Applications of Chaos and Fractals—International State-of-the-Art Seminar (London)* (Ch 3)

Allen J S, Samelson R M and Newberger P A 1991 Chaos in a model of forced quasi-geostrophic flow over topography: an application of Melinkov's method *J. Fluid Mech.* **226** 511–47 (Ch 6)

Allen M, Brown G J and Miles N J 1995 Measurement of boundary fractal dimensions: review of current techniques *Powder Technol.* **84** 1–14 (Ch 3)

Alvarez-Ramirez J 1994 Nonlinear feedback for controlling the Lorenz equation *Phys. Rev.* E **50** 2339–42 (Ch 6)

Appleby S 1995 Estimating the cost of a telecommunications network using the fractal structure of the human population distribution *IEE Proc. Commun.* **142** 172–8 (Ch 3)

Arecchi F T and Harrison R G (eds) 1987 *Instabilities and Chaos in Quantum Optics* (Berlin: Springer) (Ch 6)

Aref H and El Naschie M S (eds) 1995 *Chaos Applied to Fluid Mixing* (Oxford: Pergamon, Elsevier) (Ch 6)

Aref H, Jones S W, Mofina S and Zawadzki I 1989 Vortices, kinematics and chaos *Physica* D **37** 423–40 (Ch 6)

Argoul F, Arneodo A, Elezgaray J, Grasseau G and Murenzi R 1989 Wavelet transform of fractal aggregates *Phys. Lett.* **135A** 327–36 (Ch 4)

Arneodo A, Coullet P, Tresser C, Libchaber A, Maurer J and d'Humières D 1983 On the observation of an uncompleted cascade in a Rayleigh–Bénard experiment *Physica* D **6** 385–92 (Ch 6)

Arneodo A, d'Auberton-Carafa Y, Bacry E, Graves P V, Muzy J F and Thermes C 1996 Wavelet based fractal analysis of DNA sequences *Physica* D **96** 291–320 (Ch 4)

Artuso R, Cvitanovic P and Casati G (eds) 1991 *Chaos, Order, and Patterns (NATO ASI Series, 280)* (New York: Plenum) (Ch 6)

Asfar K R and Masoud K K 1992 On the period-doubling bifurcations in the Duffing's oscillator with negative linear stiffness *Trans. ASME, J. Vib. Acoust.* **114** 489–94 (Ch 6)

Aubry N, Holmes P, Lumley J L and Stone E 1988 The dynamics of coherent structures in the wall region of a turbulent boundary layer *J. Fluid Mech.* **192** 115–73 (Ch 6)

Auerbach D, Grebogi C, Ott E and Yorke J A 1992 Controlling chaos in high dimensional systems *Phys. Rev. Lett.* **69** 3479–82 (Ch 7)

Avnir D 1989 *The Fractal Approach to Heterogeneous Chemistry: Surfaces, Colloids, Polymers* (Chichester: Wiley) (Ch 3)

Awrejcewicz J 1991 *Bifurcation and Chaos in Coupled Oscillators* (Singapore: World Scientific) (Ch 6)

Azevedo A and Rezende S M 1991 Controlling chaos in spin-wave instabilities *Phys. Rev. Lett.* **66** 1342–45 (Ch 7)

Badii R, Broggi G, Derighetti B, Ravani M, Ciliberto S, Politi A and Rubio M A 1988 Dimension increase in filtered chaotic signals *Phys. Rev. Lett.* **60** 979–82 (Ch 7)

Bai-Lin H (ed) 1984 *Chaos* (Singapore: World Scientific) (Ch 6)

——1987 *Directions in Chaos: Volume 1* (Singapore: World Scientific) (Ch 6)

——1989 *Elementary Symbolic Dynamics and Chaos in Dissipative Systems* (Singapore: World Scientific) (Ch 6)

——1990 *Directions in Chaos. Volume 3: Experimental Study and Characterization of Chaos* (Singapore: World Scientific) (Ch 7)

Bak P and Chen K 1989 The physics of fractals *Physica* D **38** 5–12 (Ch 6)

Bak P, Chen K and Tang C 1990 A forest-fire model and some thoughts on turbulence *Phys. Lett.* **147A** 297–300 (Ch 6)

Baker G L and Gollub J P 1990 *Chaotic Dynamics: an Introduction* (Cambridge: Cambridge University Press) (Ch 6)

Balazs L, Fleury V, Duclos F and Van Herpen A 1996 Fractal growth of silicon-rich domains during annealing of aluminium thin films deposited on silica *Phys. Rev.* E **54** 559–604 (Ch 4)

Balkin S D, Golebiewski E L and Reiter C A 1994 Chaos and elliptic curves *Comput. Graph.* **18** 113–7 (Ch 5)

Bapat C N 1995 Duffing oscillator under periodic impulses *J. Sound Vib.* **179** 725–32 (Ch 6)

Barabási A-L and Stanley H E 1995 *Fractal Concepts in Surface Growth* (Cambridge: Cambridge University Press) (Ch 3, Ch 4)

Barenblatt G I, Iooss G and Joseph D D (eds) 1983 *Nonlinear Dynamics and Turbulence* (Boston, MA: Pitman) (Ch 6)

Barlow D and Gowan R 1988 *Chaos* An Independent Communications Associates, World's Edge Films Production for Channel 4 (UK) (Ch 1)

Barndorff-Nielsen O E, Jensen J L and Kendall W S (eds) 1993 *Networks and Chaos—Statistical and Probabilistic Aspects* (London: Chapman and Hall) (Ch 6)

Barnett W and Chen P 1988 Deterministic chaos and fractal attractors as tools for nonparametric dynamical econometric inference: with an application to the divisia monetary aggregates *Math. Comput. Modelling* **10** 275–96 (Ch 6)

Barnett W A, Geweke J and Shell K (eds) 1989 *Economic Complexity: Chaos, Sunspots, Bubbles, and Nonlinearity, Proc. 4th Int. Symp. in Economic Theory and Econometrics* (Cambridge: Cambridge University Press) (Ch 6)

Barnsley M 1988 *Fractals Everywhere* (Boston, MA: Academic) (Ch 5)

Barnsley M F and Demko S G 1986 *Chaotic Dynamics and Fractals* (San Diego, CA: Academic) (Ch 5)

Bartoli F, Philippy R, Doirisse M, Niquet S and Dubuit M 1991 Structure and self-similarity in silty and sandy soils: the fractal approach *J. Soil Science* **42** 167–85 (Ch 3, Ch 4)

Batty M 1995 New ways of looking at cities *Nature* **377** 574 (Ch 3)

Batty M and Longley P 1986 Fractal-based description of urban form *Papers in Planning Research* **95** University of Wales Institute of Science and Technology, Department of Town Planning (Ch 3)

Bau H H, Bertram L A and Korpela S A (eds) 1992 *Bifurcation Phenomena and Chaos in Thermal Convection* (New York: American Society of Mechanical Engineers) (Ch 6)

Bazant Z P 1995 Scaling of quasi-brittle fracture and the fractal question *Trans. ASME: J. Eng. Mater. Technol.* **117** 361–7 (Ch 3)

Beck C 1994 Chaotic cascade model for turbulent velocity distributions *Phys. Rev.* E **49** 3641–52 (Ch 6)

Becker K-H and Dörfler M 1989 *Dynamical Systems and Fractals: Computer Graphics Experiments in Pascal* (Cambridge: Cambridge University Press) (Ch 5)

Beckmann P E 1995 On the problem of visualizing point distributions in high dimensional spaces *Comput. Graph.* **19** 617–29 (Ch 7)

Bélair J, Glass L, an der Heiden U and Milton J 1995 Dynamical disease: identification, temporal aspects and treatment strategies *Chaos* **5** 1–7 (This is the introductory paper in a special focus issue of the journal on 'dynamical disease: mathematical analysis of human illness') (Ch 6)

Ben-Jacob E, Schochet O, Tenenbaum A, Cohen I, Czirok A and Vicsek T 1994 Generic modelling of cooperative growth patterns in bacterial colonies *Nature* **368** 46–9 (Ch 4)

Ben-Mizrachi A, Procaccia I and Grassberger P 1984 Characterization of experimental (noisy) strange attractors *Phys. Rev.* A **29** 975–7 (Ch 7)

Benzi R, Paladin G, Parisi G and Vulpiani A 1984 On the multifractal nature of fully developed turbulence and chaotic systems *J. Phys. A: Math. Gen.* **17** 3521–31 (Ch 6)

Berge P, Pomeau Y and Vidal C 1984 *Order Within Chaos* (New York: Wiley) (Ch 6)

Berry M V, Percival I C and Weiss N O 1987 *Dynamical Chaos: Proc. R. Soc. Discussion Meeting (1987)* (London: The Royal Society) (Ch 6)

Bessant C 1993 *Computers and Chaos* (Wilmslow: Sigma) (Ch 1)

Bhavsar V C, Gujar U G and Vangala N 1993 Vectorization of generation of fractals from $z \leftarrow z^2 + c$ on IBM 3090/180VF *Comput. Graph.* **17** 169–74 (Ch 5)

Binder B 1994 Fractal Brownian motion and local 'fracton' metric obtained by proper coordinate transformation of ordinary Brownian motion *Phys. Lett.* **196A** 213–6 (Ch 4)

Boldrighini C and Franceschini V 1979 A five-dimensional truncation of the plane incompressible Navier–Stokes equations *Commun. Math. Phys.* **64** 159–70 (Ch 6)

Bolz A, Fröhlich R, Schmidt K and Schaldach M 1995 Effect of smooth, porous and fractal surface structure on the properties of an interface *J. Mater. Sci.: Mater. Medic.* **6** 844–8 (Ch 3)

Borcherds P H and Mc Cauley G P 1993 The digital tent map and the trapezoidal map *Chaos, Solitons Fractals* **3** 451–66 (Ch 5)

Borgas M S 1993 Self-similarity and multifractals in turbulence *Phys. Fluids* A **5** 3181–5 (Ch 6)

——1993 The multifractal Lagrangian nature of turbulence *Phil. Trans. R. Soc.* A **342** 379–411 (Ch 6)

Bouchaud J P, Bouchaud E, Lapasset G and Planès J 1993 Models of fractal cracks *Phys. Rev. Lett.* **71** 2240–3 (Ch 3)

Bovill C 1996 *Fractal Geometry in Architecture and Design* (Boston, MA: Birkhäuser) (Ch 3)

Bradley E 1995 Causes and effects of chaos *Comput. Graph.* **19** 755–78 (Ch 6)

Brandstäter A, Swift J, Swinney H L, Wolf A, Farmer J D, Jen E and Crutchfield P J 1983 Low-dimensional chaos in a hydrodynamic system *Phys. Rev. Lett.* **51** 1442–5 (Ch 6, Ch 7)

Brandstäter A and Swinney H L 1987 Strange attractors in weakly turbulent Couette–Taylor flow *Phys. Rev.* A **35** 2207–20 (Ch 6, Ch 7)

Brandstäter A, Swinney H L and Chapman G T 1986 Characterizing turbulent channel flow *Dimensions and Entropies in Chaotic Systems: Quantification of Complex Behaviour, Proc. Int. Workshop (Pecos River Ranch, NM, 1985)* ed G Mayer-Kress (Berlin: Springer) pp 150–7 (Ch 6)

Briggs J 1992 *Fractals: the Patterns of Chaos: Discovering a New Aesthetic of Art, Science, and Nature* (London: Thames and Hudson) (Ch 3)

Briggs J and Peat F D 1989 *Turbulent Mirror* (New York: Harper and Row (Ch 1)

Brindley J 1986 Spatio-temporal behaviour of flows with circular constraints *Physica* D **23** 240–5 (Ch 6)

Broomhead D S, Huke J P and Muldoon M R 1992 Linear filters and non-linear systems *J. R. Stat. Soc.* B **54** 373–82 (Ch 7)

Broomhead D S and King G P 1986 Extracting qualitative dynamics from experimental data *Physica* D **20** 217–36 (Ch 7)

Brown R D, Addison P and Chan A H C 1994 Chaos in the unbalance response of journal bearings *Nonlinear Dynamics* **5** 421–32 (Ch 6)

Brunsden V, Cortell J and Holmes P J 1989 Power spectra of chaotic vibrations of a buckled beam *J. Sound Vib.* **130** 1–25 (Ch 6)

Bryant P, Brown R and Abarbanel H D I 1990 Lyapunov exponents from observed time series *Phys. Rev. Lett.* **65** 1523–6 (Ch 7)

Buldyrev S V, Goldberger A L, Havlin S, Peng C-K, Simons M, Sciortino F and Stanley H E 1993 Long-range power-law correlations in DNA *Phys. Rev. Lett.* **71** 1776 (Ch 4)

Bunde A and Havlin S (eds) 1991 *Fractals and Disordered Systems* (Berlin: Springer) (Ch 3)

——1994 *Fractals in Science* (Berlin: Springer) (Ch 3)

Burioni R and Cassi D 1994 Fractals without anomolous diffusion *Phys. Rev.* E **49** 1785–7 (Ch 4)

Burton T D and Anderson M 1989 On asymptotic behaviour in cascaded chaotically excited non-linear oscillators *J. Sound Vib.* **133** 353–580 (Ch 6)

Buzug T, Pawelzik K, von Stamm J and Pfister G 1994 Mutual information and global strange attractors in Taylor–Couette flow *Physica* D **72** 343–50 (Ch 7)

Buzug T, Reimers T and Pfister G 1990 Optical reconstruction of strange attractors from purely geometrical arguments *Europhys. Lett.* **13** 605–10 (Ch 7)

Cabrera J L and de la Rubia F J 1995 Numerical analysis of transient behaviour in the discrete random logistic equation with delay *Phys. Lett.* **197A** 19–24 (Ch 5)

Çambel A B 1993 *Applied Chaos Theory: a Paradigm for Complexity* (Boston, MA: Academic) (Ch 6)

Campbell D and Rose H (eds) 1983 *Order in Chaos* (Amsterdam: North-Holland) (Ch 6)

Carlson J M and Langer J S 1989 Mechanical model of an earthquake fault *Phys. Rev.* A **40** 6470–84 (Ch 6)

Casati G (ed) 1985 *Chaotic Behaviour in Quantum Systems: Theory and Applications* (New York: Plenum) (Ch 6)

Casdagli M 1989 Nonlinear prediction of chaotic time series *Physica* D **35** 335–56 (Ch 7)

Caserta F, Stanley H E, Eldred W D, Daccord G, Hausman R E and Nittman J 1990 Physical mechanisms underlying neurite outgrowth: a quantitative analysis of neuronal shape *Phys. Rev. Lett.* **64** 95–8 (Ch 4)

Chan A H C 1996 private communication (Appendix 1)

Chandra J (ed) 1984 *Chaos in Nonlinear Dynamical Systems* (Philadelphia, PA: SIAM) (Ch 6)

Charmet J C, Roux S and Guyon E (eds) 1990 *Disorder and Fracure* (New York: Plenum) (Ch 4)

Chatterjee S and Mallik A K 1996 Bifurcations and chaos in autonomous self-excited oscillators with impact damping *J. Sound Vib.* **191** 539–62 (Ch 6)

Chen C-C, Tsai C-H and Fu C-C 1994 Rich dynamics in self-interacting Lorenz systems *Phys. Lett.* **194** 265–71 (Ch 6)

Cheng Q 1995 The perimeter–area fractal model and its application to geology *Math. Geol.* **27** 69–82 (Ch 3, Ch 4)

Chennaoui A, Pawelzik K, Liebert W, Schuster H G and Pfister G 1990 Attractor reconstruction from filtered chaotic signals *Phys. Rev.* A **41** 4151–9 (Ch 7)

Cherbit G (ed) 1991 *Fractals: Non-Integral Dimensions and Applications* (Chichester: Wiley) (Ch 3)

Chern J-L 1994 Memory-flow structures in Lorenz chaos *Phys. Rev.* E **50** 4315–8 (Ch 6)

Chevray R and Mathieu J 1993 *Topics in Fluid Mechanics* (Cambridge: Cambridge University Press) (Ch 6)

Choi H S and Lou J Y K 1991 Nonlinear behaviour of an articulated offshore loading platform *Appl. Ocean Res.* **13** 63–74 (Ch 6)

Choi S-Y and Lee E K 1995 Scaling behaviour at the onset of chaos in the logistic map driven by coloured noise *Phys. Lett.* **205A** 173–8 (Ch 5)

Christini D J and Collins J J 1996 Using chaos control and tracking to suppress a pathological nonchaotic rhythm in a cardiac model *Phys. Rev.* E **53** 49–52 (Ch 7)

Chua L O and Hasler M (eds) 1993 Special issue on chaos in nonlinear electronic circuits—Part C: Applications *IEEE Trans. Circuits Syst. Part 2: Analog and Digital Signal Processing* **40** No 10 (Ch 6)

Cioczek-Georges R and Mandelbrot B B 1995 A class of micropulses and antipersistent fractional Brownian motion *Stochast. Processes Appl.* **60** 1–18 (Ch 4)

Cohen A and Procaccia I 1985 Computing the Kolmogorov entropy from time signals of dissipative and conservative dynamical systems *Phys. Rev.* A **31** 1872–82 (Ch 7)

Coleman P H and Pietronero L 1992 The fractal structure of the universe *Phys. Rep.* **213** 311–89 (Ch 3)

Colet P and Braiman Y 1996 Control of chaos in multimode solid state lasers by the use of small periodic perturbations *Phys. Rev.* E **53** 200–6 (Ch 6)

Collins J J and De Luca C J 1994 Random walking during quiet standing *Phys. Rev. Lett.* **73** 764–7 (Ch 4)

Collins J J and Stewart I 1993 Hexapodal gaits and coupled nonlinear oscillator models *Biol. Cybernet.* **68** 287–98 (Ch 6)

Compte A 1996 Stochastic foundations of fractional dynamics *Phys. Rev.* E **53** 4191–3 (Ch 4)

Conlisk A T, Guezennec Y G and Elliot G S 1989 Chaotic motion of an array of vortices above a flat wall *Phys. Fluids* A **1** 704–17 (Ch 6)

Constantin P, Procaccia I and Sreenivasan K R 1991 Fractal geometry of isoscalar surfaces in turbulence: theory and experiments *Phys. Rev. Lett.* **67** 1739–42 (Ch 6)

Cooley J W, Lewis P A W and Welch P D 1969 The fast Fourier transform and its applications *IEEE Trans. Education* E **12** 27–34 (Ch 7)

Coppens M-O and Froment G F 1994 Diffusion and reaction in a fractal catalyst pore—III. Application to the simulation of vinyl acetate production from ethylene *Chem. Eng. Sci.* **49** 4897–907 (Ch 3)

——1995a Diffusion and reaction in a fractal catalyst pore—I. Geometrical aspects *Chem. Eng. Sci.* **50** 1013–26 (Ch 3)

——1995b Diffusion and reaction in a fractal catalyst pore—II. Diffusion and first-order reaction *Chem. Eng. Sci.* **50** 1027–39 (Ch 3)

Crawford J W, Ritz K and Young I M 1993 Quantification of fungal morphology, gaseous transport and microbial dynamics in soil: an integrated framework utilising fractal geometry *Geoderma* **56** 157–72 (Ch 3, Ch 4)

Creedy J and Martin V L 1994 *Chaos and Non-Linear Models in Economics* (Aldershot: Elgar) (Ch 6)

Cross S S 1994 The application of fractal geometric analysis to microscopic images *Micron* **25** 101–13 (Ch 3)

Crutchfield J, Farmer D, Packard N, Shaw R, Jones G and Donelly R J 1980 Power spectral analysis of a dynamical system *Phys. Lett.* **76A** 1–4 (Ch 7)

Csanady G T 1973 *Turbulent Diffusion in the Environment* (Dordrecht: Reidel) (Ch 4)

Cuomo K M and Oppenheim A V 1993 Circuit implementation of synchronized chaos with applications to communications *Phys. Rev. Lett.* **71** 65–8 (Ch 6)

Cuomo K M, Oppenheim A V and Strogatz S H 1993 Synchronization of Lorenz-based chaotic circuits with applications to communications *IEEE Trans. Circuits Systems II: Analog and Digital Signal Processing* **40** 626–33 (Ch 6)

Cvitanovic P (ed) 1984 *Universality in Chaos* (Bristol: Hilger) (Ch 6)

Cvitanovic P and Myrheim J 1989 Complex universality *Commun. Math. Phys.* **121** 225–54 (Ch 5)

Dabby D S 1996 Musical variations from a chaotic mapping *Chaos* **6** 95–107 (Ch 6)

Daccord G 1987 Chemical dissolution of a porous medium by a reactive fluid *Phys. Rev. Lett.* **58** 479–82 (Ch 4)

Darbyshire A G and Broomhead D S 1996 Robust estimation of tangent maps and Liapunov spectra *Physica* D **89** 287–305 (Ch 7)

David H 1995 Two fractals based on Keplerian solids *Comput. Graph.* **19** 885–8 (Ch 2)

Degn H, Holden A V and Olsen L F (eds) 1987 *Chaos in Biological Systems (NATO ASI Series A: Life Sciences 138)* (New York: Plenum) (Ch 6)

Deissler R G 1986 Is Navier–Stokes turbulence chaotic? *Phys. Fluids* **29** 1453–57 (Ch 6)

Deissler R J and Kaneko K 1987 Velocity-dependent Lyapunov exponents as a measure of chaos for open-flow systems *Phys. Lett.* **119A** 397–402 (Ch 7)

Dekhtyaryuk E S, Zakharchenko T G, Petryna Y S and Krasnopolskaya T S 1994 Four modes competition and chaos in a shell *Chaos* **4** 637–50 (Ch 6)

Delay F, Housset-Resche H, Porel G and de Marsily G 1996 Transport in a 2-D saturated porous medium: a new method for particle tracking *Math. Geol.* **28** 45–71 (Ch 4)

Denton T A, Diamond G A, Helfant R H, Khan S and Karagueuzian H 1990 Fascinating rhythm: a primer on chaos theory and its application to cardiology *Am. Heart J.* **120** 1419–40 (Ch 6)

Devaney R L 1989 *An Introduction to Chaotic Dynamical Systems* 2nd edn (Redwood City, CA: Addison-Wesley) (Ch 6)

Devaney R L and Keen L (eds) 1989 *Chaos and Fractals: the Mathematics Behind the Computer Graphics* (Providence, RI: American Mathematical Society) (Ch 5)

Dhurandhar S V, Bhavsar V C and Gujar U G 1993 Analysis of z-plane images from $z \leftarrow z^{\alpha} + c$ for $\alpha < 0$ *Comput. Graph.* **17** 89–94 (Ch 5)

Ding M, Grebogi C, Ott E, Sauer T and Yorke J A 1993 Plateau onset for correlation dimension: when does it occur? *Phys. Rev. Lett.* **70** 3872–6 (Ch 7)

Ding M and Yang W 1995 Distribution of the first return time in fractional Brownian motion and its application to the study of on–off intermittency *Phys. Rev.* E **52** 207–13 (Ch 4)

Ditto W L, Rauseo S N and Spano M L 1990a Experimental control of chaos *Phys. Rev. Lett.* **65** 3211–4 (Ch 7)

Ditto W L, Spano M L, Savage H T, Rauseo S N, Heagy J and Ott E 1990b Experimental observation of a strange nonchaotic attractor *Phys. Rev. Lett.* **65** 533–6 (Ch 7)

Dixon S L, Steele K L and Burton R P 1996 Generation and graphical analysis of Mandelbrot and Julia sets in more than four dimensions *Comput. Graph.* **20** 451–6 (Ch 5)

Dobrushin R L and Kosuka S 1993 *Statistical Mechanics and Fractals (Lecture Notes in Mathematics 1567)* (Berlin: Springer) (Ch 3)

Doherty M F and Ottino J M 1988 Chaos in deterministic systems: strange attractors, turbulence, and applications in chemical engineering *Chem. Eng. Sci.* **43** 139–83 (Ch 6)

Dolnik M and Epstein I R 1996 Coupled chaotic chemical oscillators *Phys. Rev.* E **54** 3361–8 (Ch 6)

Douady A and Hubbard J H 1985 On the dynamics of polynomial-like mappings *Ann. Sci. École Normale Supérieure* **18** 287–343 (Ch 5)

Dowell E H and Pezeshki C 1986 On the understanding of chaos in Duffings equation including a comparison with experiment *Trans. ASME, J. Appl. Mech.* **53** 5–9 (Ch 6)

——1988 On necessary and sufficient conditions for chaos to occur in Duffing's equation: an heuristic approach *J. Sound Vib.* **121** 195–200 (Ch 6)

Du C, Satik C and Yortos Y C 1996 Percolation in a fractional Brownian motion lattice *AICh E J.* **42** 2392–5 (Ch 4)

Dvořák I and Klaschka J 1990 Modification of the Grassberger–Procaccia algorithm for estimating the correlation exponent of chaotic systems with high embedding dimension *Phys. Lett.* **145A** 225–31 (Ch 7)

Eckmann J-P, Kamphorst S O, Ruelle D and Ciliberto S 1986 Liapunov exponents from time series *Phys. Rev.* A **34** 4971–9 (Ch 7)

Eckmann J-P and Ruelle D 1985 Ergodic theory of chaos and strange attractors *Rev. Mod. Phys.* **57** 617–56 (Ch 7)

——1992 Fundamental limitations for estimating dimensions and Lyupanov exponents in dynamical systems *Physica* D **56** 185–7 (Ch 7)

Edgar G A 1990 *Measure, Topology and Fractal Geometry* (New York: Springer) (Ch 3)

Einstein A 1926 *Investigations on the Theory of the Brownian Movement* (London: Methuen) (Ch 4)

Eisenberg E, Havlin S and Weiss G H 1994 Fluctuations of the probability density of diffusing particles for different realizations of a random medium *Phys. Rev. Lett.*, **72** 2827–30 (Ch 4)

Elgar S, Van Atta C W and Gharib M 1989 Bispectral analysis of ordered and chaotic vortex shedding from vibrating cylinders at low Reynolds numbers *Physica* D **39** 281–6 (Ch 6)

El Naschie M S 1990 *Stress, Stability and Chaos in Structural Engineering: An Energy Approach* (London: McGraw-Hill) (Ch 6)

Encarnação J L, Peitgen H-O, Sakas G and Englert G (eds) 1992 *Fractal Geometry and Computer Graphics* (Berlin: Springer) (Ch 4)

Erramilli A, Singh R P and Pruthi P 1995 An application of deterministic chaotic maps to model packet traffic *Queueing Systems* **20** 171–206 (Ch 5)

Falconer K J 1985 *The Geometry of Fractal Sets* (Cambridge: Cambridge University Press) (Ch 3)

——1990 *Fractal Geometry: Mathematical Foundations and Applications* (Chichester: Wiley) (Ch 3)

Falsone G and Elishakoff I 1994 Modified stochastic linearization technique for colored noise excitation of Duffing oscillator *Int. J. Non-Linear Mech.* **29** 65–9 (Ch 6)

Family F, Masters B R and Platt D E 1989 Fractal pattern formation in human retinal vessels *Physica* D **38** 98–103 (Ch 3)

Fan L T, Neogi D and Yashima M 1991 *Elementary Introduction to Spatial and Temporal Fractals (Lecture Notes in Chemistry 55)* (Berlin: Springer) (Ch 3)

Farmer J D 1981 Spectral broadening of period-doubling bifurcation sequences *Phys. Rev. Lett.* **47** 179–82 (Ch 7)

——1982 Chaotic attractors of an infinite dimensional dynamical system *Physica* D **4** 366–93 (Ch 6, Ch 7)

Farmer D, Crutchfield J, Froehling H, Packard N and Shaw R 1980 Power spectra and mixing properties of strange attractors *Ann. NY Acad. Sci.* **357** 453–72 (Ch 7)

Farmer J D, Ott E and Yorke J A 1983 The dimension of chaotic attractors *Physica* D **7** 153–80 (Ch 7)

Feder J 1988 *Fractals* (New York: Plenum) (Ch 3, Ch 4)

Feder J, Hinrichsen E L, Maloy K J and Jossang T 1989 Geometrical crossover and self-similarity of DLA and viscous fingering clusters *Physica* D **38** 104–11 (Ch 4)

Feigenbaum M J 1978 Quantitative universality for a class of nonlinear transformations *J. Stat. Phys.* **19** 25–52 (Ch 5, Ch 6)

——1980 *Universal Behaviour in Nonlinear Systems* **1** (Los Alamos Science) 4–27 (Ch 5, Ch 6)

Femat R, Alvarez-Ramirez J and Zarazua M 1996 Chaotic behaviour from a human biological signal *Phys. Lett.* **214A** 175–9 (Ch 6)

Ferer M and Smith D H 1994 Dynamics of growing interfaces from the simulation of unstable flow in random media *Phys. Rev.* E **49** 4114–20 (Ch 4)

Fischer P and Smith W R (eds) 1985 *Chaos, Fractals, and Dynamics* (New York: Dekker) (Ch 6)

Fisher G V 1993 An introduction to chaos theory and some haematological applications *Comparative Haematol. Int.* **3** 43–51 (Ch 6)

Fleischmann M, Tildesley D J and Ball R C (eds) 1989 *Fractals in the Natural Sciences* (Princeton, NJ: Princeton University Press) (Ch 3)

Flohr P and Olivari D 1994 Fractal and multifractal characteristics of a scalar dispersed in a turbulent jet *Physica* D **76** 278–90 (Ch 3)

Flook A G 1982 Fractal dimensions: their evaluation and their significance in stereological measurements *Acta Stereol.* **1** 79–87 (Ch 3)

Foias C and Treve Y M 1981 Minimum number of modes for the approximation of the solutions of the Navier–Stokes equations in two and three dimensions *Phys. Lett.* **85A** 35–7 (Ch 6)

Fournier A, Fussell D and Carpenter L 1982 Computer rendering of stochastic models *Commun. ACM* **25** 371–84 (Ch 4)

Fraedrich K and Wang R 1993 Estimating the correlation dimension of an attractor from noisy and small datasets based on re-embedding *Physica* D **65** 373–98 (Ch 7)

Frame M, Philip A GD and Robucci A 1992 A new scaling along the spike of the Mandelbrot set *Comput. Graph.* **16** 223–34 (Ch 5)

Frame M and Robertson J 1992 A generalized Mandelbrot set and the role of critical points *Comput. Graph.* **16** 35–40 (Ch 5)

Franceschini V and Tebaldi C 1979 Sequences of infinite bifurcations and turbulence in a five-mode truncation of the Navier–Stokes equations *J. Stat. Phys.* **21** 707–26 (Ch 6)

Fraser A M 1989 Reconstructing attractors from scalar time series: a comparison of singular system and redundancy criteria *Physica* D **34** 391–404 (Ch 7)

Fraser A M and Swinney H L 1986 Independent coordinates for strange attractors from mutual information *Phys. Rev.* A **33** 1134–40 (Ch 7)

Frisch U 1995 *Turbulence: The Legacy of A N Kolmogorov* (Cambridge: Cambridge University Press) (Ch 6)

Frisch U and Orszag S A 1990 Turbulence: challenges for theory and experiment *Phys. Today* **43** 24–32 (Ch 6)

Frisch U and Vergassola M 1991 A prediction of the multifractal model: the intermediate dissipation range *Europhys. Lett.* **14** 439–44 (Ch 6)

Froyland J 1992 *Introduction to Chaos and Coherence* (Bristol: Institute of Physics Publishing) (Ch 5)

Froyland J and Alfsen K H 1984 Lyapunov-exponent spectra for the Lorenz model *Phys. Rev.* A **29** 2928–31 (Ch 7)

Gade P M and Amritkar R E 1994 Wavelength-doubling bifurcations in one-dimensional coupled logistic maps *Phys. Rev.* E **49** 2617–22 (Ch 5)

Gallas J A C 1993 Structure of the parameter space of the Hénon map *Phys. Rev. Lett.* **70** 2714–7 (Ch 5)

Garrido L (ed) 1983 Dynamical systems and chaos *Lecture Notes in Physics* (Berlin: Springer) (Ch 6)

Gay J-C (ed) 1992 *Irregular Atomic Systems and Quantum Chaos* (Philadelphia, PA: Gordon and Breach) (Ch 6)

Ge Z-M, Chen H-K and Chen H-H 1996 The regular and chaotic motions of a symmetric heavy gyroscope with harmonic excitation *J. Sound Vib.* **198** 131–47 (Ch 6)

Gershenfeld N A 1992 Dimension measurement on high dimensional systems *Physica* D **55** 135–54 (Ch 7)

Giaever I and Keese C R 1989 Fractal motion of mammalian cells *Physica* D **38** 128–33 (Ch 4)

Gibbs H M, Hopf F A, Kaplan D L and Shoemaker R L 1981 Observation of chaos in optical bistability *Phys. Rev. Lett.* **46** 474–7 (Ch 6)

Giglio M, Musazzi S and Perini U 1981 Transition to chaotic behaviour via a reproducible sequence of period doubling bifurcations *Phys. Rev. Lett.* **47** 243–6 (Ch 6)

Gilbert W J 1994 Newton's method for multiple roots *Comput. Graph.* **18** 227–9 (Ch 5)

Gioggia R S and Abraham N B 1983 Routes to chaotic output from a single-mode, dc-excited laser *Phys. Rev. Lett.* **51** 650–3 (Ch 6)

Glass L and Mackey M C 1988 *From Clocks to Chaos: the Rhythms of Life* (Princeton, NJ: Princeton University Press) (Ch 6)

Gleick J 1987 *Chaos: Making a New Science* (London: Cardinal) (Ch 1)

Glendinning P 1994 *Stability, Instability and Chaos: an Introduction to the Theory of Nonlinear Differential Equations* (Cambridge: Cambridge University Press) (Ch 6)

Gollub J P and Benson S V 1980 Many routes to turbulent convection *J. Fluid Mech.* **100** Part 3, 449–70 (Ch 6)

Gollub J P and Swinney H L 1975 Onset of turbulence in a rotating fluid *Phys. Rev. Lett.* **35** 927–30 (Ch 6)

Goodchild M F 1988 Lakes on fractal surfaces: a null hypothesis for lake-rich landscapes *Math. Geol.* **20** 615–30 (Ch 3)

Goodwin R M 1990 *Chaotic Economic Dynamics* (Oxford: Clarendon) (Ch 6)

Gottwald J A, Virgin L N and Dowell E H 1992 Experimental mimicry of Duffing's equation *J. Sound Vib.* **158** 447–67 (Ch 6)

Gourley P L, Tigges C P, Schneider R P, Brennan T M, Hammons B E and McDonald A E 1993 Optical properties of fractal quantum wells *Appl. Phys. Lett.* **62** 1736–8 (Ch 3)

Grabec I 1986 Chaos generated by the cutting process *Phys. Lett.* **117A** 384–6 (Ch 6)

Grassberger P 1981 On the Hausdorff dimension of fractal attractors *J. Stat. Phys.* **26** 173–9 (Ch 5)

——1983a On the fractal dimension of the Henon attractor *Phys. Lett.* **97A** 224–6 (Ch 5)

——1983b Generalized dimensions of strange attractors *Phys. Lett.* **97A** 227–30 (Ch 7)

——1986 Do climatic attractors exist? *Nature* **323** 609–12 (Ch 7)

——1987 Reply to C Nicolis and G Nicolis *Nature* **326** 524 (Ch 7)

——1990 An optimized box-assisted algorithm for fractal dimensions *Phys. Lett.* **148A** 63–8 (Ch 7)

——1993 On efficient box counting algorithms *Int. J. Mod. Phys.* C **4** 515–23 (Ch 7)

Grassberger P, Hegger R, Kantz H, Schaffrath C and Schreiber T 1993 On noise reduction methods for chaotic data *Chaos* **3** 127–41 (Ch 7)

Grassberger P and Procaccia I 1983a Characterisation of strange attractors *Phys. Rev. Lett.* **50** 346–9 (Ch 7)

——1983b Measuring the strangeness of strange attractors *Physica* D **9** 189–208 (Ch 7)

——1983c Estimation of the Kolmogorov entropy from a chaotic signal *Phys. Rev.* A **28** 2591–3 (Ch 7)

Grebogi C, Ott E, Pelikan S and Yorke J A 1984 Strange attractors that are not chaotic *Physica* D **13** 261–8 (Ch 7)

Grey F and Kjems J K 1989 Aggregates, broccoli and cauliflower *Physica* D **38** 154–9 (Ch 3)

Grier D, Ben-Jacob E, Clarke R and Sander L M 1986 Morphology and microstructure in electrochemical deposition of zinc *Phys. Rev. Lett.* **56** 1264–7 (Ch 4)

Gripenberg G and Norros I 1996 On the prediction of fractional Brownian motion *J. Appl. Prob.* **33** 400–10 (Ch 4)

Guckenheimer J 1986 Strange attractors in fluids: another view *Ann. Rev. Fluid Mech.* **18** 15–31 (Ch 6)

Guckenheimer J and Buzyna G 1983 Dimension measurements for geostrophic turbulence *Phys. Rev. Lett.* 1438–41 (Ch 7)

Guckenheimer J and Holmes P 1983 *Nonlinear Oscillations, Dynamical Systems, and Bifurcations of Vector Fields* (New York: Springer) (Ch 6)

Gujar U G, Bhavsar V C and Vangala N 1992 Fractal images from $z \leftarrow z^\alpha + c$ in the complex z-plane *Comput. Graph.* **16** 45–9 (Ch 5)

Gurzadyan V G and Kocharyan A A 1991 On the nature of the fractal structure of the universe *Europhys. Lett.* **15** 801–4 (Ch 3)

Gutierrez J M, Iglesias A and Rodriguez M A 1993 Logistic map driven by dichotomous noise *Phys. Rev.* E **48** 2507–13 (Ch 5)

Gutzwiller M C 1990 *Chaos in Classical and Quantum Systems* (New York: Springer) (Ch 6)

Hagedorn P 1981 *Non-Linear Oscillations (Oxford Engineering Science Series)* (Oxford: Clarendon) (Ch 6)

Haken H (ed) 1981 *Chaos and Order in Nature: Proc. Int. Symp. on Synergetics (Schloss Elmau, 1981)* (Berlin: Springer) (Ch 6)

Hale J K and Kocak H 1991 *Dynamics and Bifurcations* (New York: Springer) (Ch 6)

Hall N 1992 *The New Scientist Guide to Chaos* (London: Penguin) (Ch 1)

Hall P and Wood A 1993 On the performance of box-counting estimators of fractal dimension *Biometrika* **80** 246–52 (Ch 7)

Halsey T C, Jensen M H, Kadanoff L P, Procaccia I and Shraiman B I 1986 Fractal measures and their singularities: the characterization of strange sets *Phys. Rev.* A **33** 1141–51 (Ch 7)

Hamburger D, Biham O and Avnir D 1996 Apparent fractality emerging from models of random distributions *Phys. Rev.* E **53** 3342–58 (Ch 3)

Harrison A 1995 *Fractals in Chemisty* (Oxford: Oxford University Press) (Ch 3)

Harrison R G and Biswas D J 1986 Chaos in light *Nature* **321** 394–401 (Ch 6)

Harrison R G, Lu W, Ditto W, Pecora L, Spano M and Vohra S (eds) 1996 *Proc. 3rd Experimental Chaos Conf. (Edinburgh, 1995)* (Singapore: World Scientific) (Ch 6)

Hartwich K and Fick E 1993 Hopf bifurcations in the logistic map with oscillating memory *Phys. Lett.* **177A** 305–10 (Ch 5)

Hasegawa M, Liu J and Konishi Y 1993 Characterization of engineering surfaces by fractal analysis *Int. J. Japan Soc. Precision Eng.* **27** 192–6 (Ch 4)

Hassell M P, Comins H N and May R M 1991 Spatial structure and chaos in insect population dynamics *Nature* **353** 255–8 (Ch 6)

Hastings H M and Sugihara G 1993 *Fractals: a User's Guide for the Natural Sciences* (Oxford: Oxford University Press) (Ch 3)

Havlin S and Ben Avraham B 1987 Diffusion in disordered media *Adv. Phys.* **36** 695–798 (Ch 4)

Havlin S, Buldyrev S V, Goldberger A L, Mantegna R N, Ossadnik S M, Peng C-K, Simons M and Stanley H E 1995 Fractals in biology and medicine *Chaos, Solitons Fractals* 171–201 (Ch 3)

Heagy J F, Carroll T L and Pecora L M 1994 Synchronous chaos in coupled oscillator systems *Phys. Rev.* E **50** 1874–85 (Ch 6)

Heagy J F and Hammel S M 1994 The birth of strange nonchaotic attractors *Physica* D **70** 140–53 (Ch 7)

Helleman R H G (ed) 1980 *Ann. NY Acad. Sci.* **357** (New York: New York Academy of Sciences) (Ch 6)

Hénon M 1976 A two-dimensional mapping with a strange attractor *Commun. Math. Phys.* **50** 69–77 (Ch 5)

Hentschel H GE and Procaccia I 1983 The infinite number of generalized dimensions of fractals and strange attractors *Physica* D **8** 435–44 (Ch 7)

——1984 Relative diffusion in turbulent media: the fractal dimension of clouds *Phys. Rev.* A **29** 1461–70 (Ch 3, Ch 4)

Herzfeld U C, Kim I I and Orcutt J A 1995 Is the ocean floor a fractal? *Math. Geol.* **27** 421–62 (Ch 3, Ch 4)

Higuchi T 1990 Relationship between the fractal dimension and the power law index for a time series: a numerical investigation *Physica* D **46** 244–64 (Ch 4)

Hilborn R C 1994 *Chaos and Non-Linear Dynamics: an Introduction for Scientists and Engineers* (Oxford: Oxford University Press) (Ch 5, Ch 7)

Hirst B and Mandelbrot B 1994 *Fractal Landscapes from the Real World* (Manchester: Cornerhouse Publications) (Ch 3)

Holden A V (ed) 1986 *Chaos* (Manchester: Manchester University Press) (Ch 6)

Holmes P 1979 A nonlinear oscillator with a strange attractor *Phil. Trans. R. Soc.* **292** 419–48 (Ch 6)

Holmgren R A 1994 *A First Course in Discrete Dynamical Systems* (New York: Springer) (Ch 5)

Hommes C 1989 Periodic, quasi-periodic and chaotic dynamics in a simple piecewise linear non-Walsarian macromodel *University of Groningen Institute of Economic Faculty of Economics, Research Memorandum 322* (Ch 6)

Hoppensteadt F C (ed) 1979 *Nonlinear Oscillations in Biology* (Providence, RI: American Mathematical Society) (Ch 6)

Hou X-J, Gilmore R, Mindlin G B and Solari H G 1990 An efficient algorithm for fast $O(N^* \ln(N))$ box counting *Phys. Lett.* **151A** 43–6 (Ch 7)

Hsü K J and Hsü A J 1992 Fractal geometry of music: from birdsongs to Bach *Proc. Applications of Chaos and Fractals—International State-of-the-Art Seminar (London, 1992)* (Ch 3)

Huang B-N and Yang C W 1995 The fractal structure in multinational stock returns *Appl. Econom. Lett.* **2** 67–71 (Ch 3)

Huang J and Turcotte D L 1990a Fractal image analysis: application to the topography of Oregon and synthetic images *J. Opt. Soc. Am.* A **7** 1124–30 (Ch 4)

——1990b Are earthquakes an example of deterministic chaos? *Geophys. Res. Lett.* **17** 223–6 (Ch 6)

Huang Y-N and Huang Y-D 1989 On the transition to turbulence in pipe flow *Physica* D **37** 153–9 (Ch 6)

Hudson J L, Mankin J, Mc Cullough J and Lamba P 1981 Experiments on chaos in a continuous stirred reactor *Nonlinear Phenomena in Chemical Dynamics: Proc. Int. Conf. (Bordeaux, 1981)* ed C Vidal and A Pacault (Berlin: Springer) pp 44–8 (Ch 6)

Hunter J R, Craig P D and Philips H E 1993 On the use of random walk models with spatially variable diffusivity *J. Comput. Phys.* **106** 366–76 (Ch 4)

Hurst H E 1951 Long term storage capacity of reservoirs *Trans. Am. Soc. Civil Eng.* **116** 770–808 (Ch 4)

——1956 Methods of using long-term storage in reservoirs *Proc. Inst. Civil Eng.: Part 1. General* **5** 519–90 (Ch 4)

Hwang S C and Yang H S 1993 Discrete approximation of the Koch curve *Comput. Graph.* **17** 95–102 (Ch 2)

IEEE 1993 Special section on fractals in electrical engineering *Proc. IEEE* **81** 1423–33 (Ch 3)

Ikeda K, Daido H and Akimoto O 1980 Optical turbulence: chaotic behaviour of transmitted light from a ring cavity *Phys. Rev. Lett.* **45** 709–12 (Ch 6)

Ingraham R L 1992 *A Survey of Nonlinear Dynamics (Chaos Theory)* (Singapore: World Scientific) (Ch 6)

Iooss G, Helleman R H G and Stora R (eds) 1983 *Chaotic Behaviour of Deterministic Systems* (Amsterdam: North Holland) (Ch 6)

Ishii H, Fujisaka H and Inoue M 1986 Breakdown of chaos symmetry and intermittency in the double-well potential system *Phys. Lett.* **116A** 257–63 (Ch 6)

Issa M A and Hammad A M 1994 Assessment and evaluation of fractal dimension of concrete fracture surface digitized images *Cement Concrete Res.* **24** 325–34 (Ch 3, Ch 4)

Jacquin A 1994 An introduction to fractals and their applications in electrical engineering *J. Franklin Inst.* B **331** 659–80 (Ch 3)

Jarrett D and Xiaoyan Z 1992 The dynamic behaviour of road traffic flow: stability or chaos? *Applications of Chaos and Fractals—Int. State-of-the-Art Seminar (London, 1992)* (Ch 6)

Johnson L W and Riess R D 1982 *Numerical Analysis* 2nd edn (Reading, MA: Addison-Wesley) (Ch 6)

Jordan D W and Smith P 1987 *Nonlinear Ordinary Differential Equations (Oxford Applied Mathematics and Computing Science Series)* 2nd edn (Oxford: Clarendon) (Ch 6)

Jun-Zheng P and Zhi-Xiong L 1994 Fractal dimension of soil strengths for paddy fields in China *J. Terramech.* **31** 1–9 (Ch 4)

Kaandorp J A 1991 Modelling growth forms of sponges with fractal techniques *Fractals and Chaos* ed A J Crilly, R A Earnshaw and H Jones (New York: Springer) pp 71–88 (Ch 3)

Kadanoff L P 1986 Fractals: where's the physics? *Phys. Today* **39** 6–7 (Ch 3)

——1993 *From Order to Chaos. Essays: Critical, Chaotic and Otherwise* (Singapore: World Scientific) (Ch 6)

Kalia R K and Vashishta P (eds) 1982 *Melting, Localization, and Chaos: Proc. 9th Midwest Solid State Theory Symp. (Argonne, IL, 1981)* (New York: North-Holland) (Ch 6)

Kaneko K 1989 Pattern dynamics in spatiotemporal chaos *Physica* D **34** 1–41 (Ch 6)

——1990 Globally coupled chaos violates the law of large numbers but not the central limit theorem *Phys. Rev. Lett.* **65** 1391–94 (Ch 6)

——1992 Overview of coupled map lattices *Chaos* **2** 279–82 (Ch 6) (This is the introductory paper in a special focus issue of the journal on coupled map lattices.)

Kapitaniak T 1986 Chaotic distribution of non-linear systems perturbed by random noise *Phys. Lett.* **116A** 251–4 (Ch 6)

Kapitaniak T 1988a Combined bifurcations and transition to chaos in a non-linear oscillator with two external periodic forces *J. Sound Vib.* **121** 259–68 (Ch 6)

——1988b Chaos in a noisy mechanical system with stress relaxation *J. Sound Vib.* **123** 391–6 (Ch 7)

——(ed) 1992 *Chaotic Oscillators: Theory and Applications* (Singapore: World Scientific) (Ch 6)

Kaplan D T and Glass L 1992 Direct test for determinism in a time series *Phys. Rev. Lett.* **68** 427–30 (Ch 7)

Kaplan D and Glass L 1995 *Understanding Nonlinear Dynamics* (New York: Springer) (Ch 6)

Kaplan J L and Yorke J A 1979 *Chaotic Behaviour of Mulitdimensional Difference Equations (Lecture Notes in Mathematics)* (Berlin: Springer) pp 204–27 (Ch 7)

Kaplan L M and Kuo C-C J 1995 Texture segmentation via Haar fractal feature estimation *J. Visual Commun. Image Representation* **6** 387–400 (Ch 4)

——1996 An improved method for 2-D self-similar image synthesis *IEEE Trans. Image Processing* **5** 754–61 (Ch 4)

Kawabe T and Kondo Y 1993 Bifurcations of the complex dynamical system $Z_{n+1} = \ln(Z_n) + C$ *J. Phys. Soc. Japan* **62** 497–505 (Ch 5)

Kaye B H 1994 *A Random Walk through Fractal Dimensions* 2nd edn (Weinheim: VCH) (Ch 3)

Keipes M, Ries F and Dicato M 1993 Of the British coastline and the interest of fractals in medicine *Biomed. Pharmacother.* **47** 409–15 (Ch 3)

Kennel M B, Brown R and Abarbanel H D I 1992 Determining embedding dimension for phase-space reconstruction using a geometrical construction *Phys. Rev.* A **45** 3403–11 (Ch 7)

Keolian R, Turkevich L A, Putterman S J, Rudnick I and Rudnick J A 1981 Subharmonic sequences in the Faraday experiment: departures from period doubling *Phys. Rev. Lett.* **47** 1133–6 (Ch 6)

Kilias T, Kelber K, Mögel A and Schwarz W 1995 Electronic chaos generators—design and applications *Int. J. Electron.* **79** 737–53 (Ch 6)

Kim J H and Stringer J (eds) 1992 *Applied Chaos* (New York: Wiley) (Ch 6)

Kobayashi M and Shimojo F 1991 Molecular dynamics studies of molten AgI: II. Fractal behaviour of diffusion trajectory *J. Phys. Soc. Japan* **60** 4076–80 (Ch 4)

Kocarev L and Parlitz U 1995 General approach for chaotic synchronization with applications to communication *Phys. Rev. Lett.* **74** 5028–31 (Ch 6)

——1996 Generalized synchronization, predictability, and equivalence of unidirectional coupled dynamical systems *Phys. Rev. Lett.* **76** 1816–9 (Ch 6)

Komori A, Baba T, Morisaki T, Kono M, Iguchi H, Nishimura K, Yamada H, Okamura S and Matsuoka K 1994 Correlation dimension and largest Lyapunov exponent for broadband edge turbulence in the compact helical system *Phys. Rev. Lett.* **73** 660–3 (Ch 7)

Korsch H J and Jodl H-J 1994 *Chaos: a Program Collection for the PC* (Berlin: Springer) (Ch 5, Ch 6)

Kostelich E J and Yorke J A 1990 Noise reduction: finding the simplest dynamical system consistent with the data *Physica* D **41** 183–96 (Ch 7)

Kotulski M 1995 Asymptotic distributions of continuous-time random walks: a probabilistic approach *J. Stat. Phys.* **81** 777–92 (Ch 4)

Krueger W M, Jost S D and Rossi K 1996 On synthesizing discrete fractional Brownian motion with applications to image processing *Graphical Models Image Processing* **58** 334–44 (Ch 4)

Krummel J R, Gardner R H, Sugihara G, O'Neill R V and Coleman P R 1987 Landscape patterns in a disturbed environment *Oikos* **48** 321–4 (Ch 3)

Kuramoto Y (ed) 1984 Chaos and statistical methods *Proc. 6th Kyoto Summer Institute (Kyoto, 1983* (Berlin: Springer) (Ch 6)

Kuznetsov Y A 1995 *Elements of Applied Bifurcation Theory* (New York: Springer) (Ch 6)

Kyriazis M 1991 Applications of chaos theory to the molecular biology of aging *Exp. Gerontol.* **26** 569–72 (Ch 6)

Labos E 1987 Chaos and neural networks *Chaos in Biological Systems (NATO ASI Series, Series A: Life Sciences 138)* ed H Degn, A V Holden and L F Olsen (New York: Plenum) pp 195–206 (Ch 6)

Lai Y-C 1996 Driving trajectories to a desirable attractor by using small control *Phys. Lett.* **221A** 375–83 (Ch 7)

Lai Y-C, Lerner D and Hayden R 1996 An upper bound for the proper delay time in chaotic time-series analysis *Phys. Lett.* **218A** 30–4 (Ch 7)

Lakshmanan M and Daniel M (eds) 1990 *Symmetries and Singularity Structures: Integrability and Chaos in Nonlinear Dynamical Systems* (Berlin: Springer) (Ch 6)

Landau L 1944 On the problem of turbulence *C. R. Acad. Sci. URSS* **44** 311 (Ch 6)

Lanford O E 1981 Strange attractors and turbulence *Hydrodynamic Instabilities and the Transition to Turbulence* ed H L Swinney and J P Gollub (Berlin: Springer) pp 7–26 (Ch 6)

Lauterborn W 1996 Nonlinear dynamics in acoustics *Acustica* **82** S46–55 (Ch 6)

Lauterborn W and Parlitz U 1988 Methods of chaos physics and their application to acoustics *J. Acoust. Soc. Am.* **84** 1975–93 (Ch 6)

Lauwerier H 1991 *Fractals: Images of Chaos* (London: Penguin) (Ch 2)

Lei T 1990 Similarity between the Mandelbrot set and Julia sets *Commun. Math. Phys.* **134** 587–617 (Ch 5)

Leung A YT and Chui S K 1995 Non-linear vibration of coupled Duffing oscillators by an improved incremental harmonic balance method *J. Sound Vib.* **181** 619–33 (Ch 6)

Leydesdorff L and van den Besselaar P 1994 *Evolutionary Economics and Chaos Theory* (London: Pinter) (Ch 6)

Li T-Y and Yorke J A 1975 Period three implies chaos *Am. Math. Monthly* **82** 985–92 (Ch 5)

Libchaber A 1984 Experiments on the onset of chaotic behaviour *J. Stat. Phys.* **34** 1047 (Ch 6)

Lichtenberg A J and Lieberman M A 1992 *Regular and Chaotic Dynamics* 2nd edn (New York: Springer) (Ch 6)

Liu H H and Molz F J 1996 Discrimination of fractional Brownian movement and fractional Gaussian noise structures in permeability and related property distributions with range analyses *Water Resources Res.* **32** 2601–5 (Ch 4)

Liu Y and Barbosa L C 1995 Period locking in coupled Lorenz systems *Phys. Lett.* **197A** 13–8 (Ch 6)

Liu Y and Leite J R R 1994 Control of Lorenz chaos *Phys. Lett.* **185A** 35–7 (Ch 7)

Liu Z, Payre G and Bourassa P 1996 Nonlinear oscillations and chaotic motions in a road vehicle system with driver steering control *Nonlinear Dynam.* **9** 281–304 (Ch 6)

Lorenz E N 1963 Deterministic nonperiodic flow *J. Atmospher. Sci.* **20** 130–41 (Ch 6)

——1993 *The Essence of Chaos* (London: UCL Press) (published in USA by University of Washington Press) (Ch 1)

Lorenz H-W 1993 *Nonlinear Dynamical Economics and Chaotic Motion* 2nd edn (Berlin: Springer) (Ch 6)

Lovejoy S 1982 Area–perimeter relationship for rain and cloud areas *Science* **216** 185–7 (Ch 3)

Lovejoy S and Mandelbrot B B 1985 Fractal properties of rain, and a fractal model *Tellus* **37A** 209–32 (Ch 3)

Lu S-Z and Hellawell A 1994 An application of fractal geometry to complex microstuctures: numerical characterization of graphite in cast irons *Acta Metall. Mater.* **42** 4035–47 (Ch 3)

Lundqvist S, March N H and Tosi M P (eds) 1988 *Order and Chaos in Nonlinear Physical Systems* (New York: Plenum) (Ch 6)

Lutzky M 1993 Counting hyperbolic components of the Mandelbrot set *Phys. Lett.* **177A** 338–40 (Ch 5)

Mackey M C and Glass L 1977 Oscillation and chaos in physiological control systems *Science* **197** 287–9 (Ch 6)

Madan R N (ed) 1993 *Chua's Circuit: a Paradigm for Chaos* (Singapore: World Scientific) (Ch 6)

Mainieri R 1993 On the equality of Hausdorff and box counting dimensions *Chaos* **3** 119–25 (Ch 3)

Malescio G 1996 Effects of noise on chaotic one-dimensional maps *Phys. Lett.* **218A** 25–9 (Ch 5)

Malinetskii G G, Potapov A B and Rakhmanov A I 1993a Limitations of delay reconstruction for chaotic dynamical systems *Phys. Rev.* E **48** 904–12 (Ch 7)

Malinetskii G G, Potapov A B, Rakhmanov A I and Rodichev E B 1993b Limitations of delay reconstruction for chaotic systems with a broad spectrum *Phys. Lett.* **179** 15–22 (Ch 7)

Mancho A M, Duarte A A and Mindlin G B 1996 Time delay embeddings and the structure of flows *Phys. Lett.* **221A** 181–6 (Ch 7)

Mandelbrot B B 1971 A fast fractional Gaussian noise generator *Water Resources Res.* **7** 543–53 (Ch 4)

——1977 *Fractals: Form, Chance and Dimension* (San Francisco, CA: Freeman) (Ch 2)

——1980 Fractal aspects of the iteration of $z \leftarrow \lambda z(1 - z)$ for complex λ and z *Ann. NY Acad. Sci.* **357** 249–59 (Ch 5)

——1982a *The Fractal Geometry of Nature* (San Francisco, CA: Freeman) (Ch 2, Ch 3, Ch 4)

——1982b Comment on computer rendering of fractal stochastic models *Commun. ACM* **25** 581–5 (Ch 4)

——1983 On the quadratic mapping $z \leftarrow z^2 - \mu$ for complex μ and z: the fractal structure of its *M* set, and scaling *Physica* D **7** 224–39 (Ch 5)

——1984a Comment on coherent structures in fluids, fractals and the fractal structure of flow singularities *Turbulence and Chaotic Phenomena in Fluids* ed T Tatsumi (Amsterdam: Elsevier) (Ch 6)

——1984b Fractals in physics: squig clusters, diffusions, fractal measures, and the unicity of fractal dimensionality *J. Stat. Phys.* **34** 895–930 (Ch 4)

Mandelbrot B B, Passoja D E and Paullay A J 1984 Fractal character of fracture surfaces of metals *Nature* **308** 721–2 (Ch 3, Ch 4)

Mandelbrot B B and Van Ness J W 1968 Fractional Brownian motions, fractional noises and applications *SIAM Rev.* **10** 422–37 (Ch 4)

Mandelbrot B B and Wallis J R 1969a Computer experiments with fractional Gaussian noises. Part 1, averages and variances *Water Resources Res.* **5** 228–41 (Ch 4)

——1969b Computer experiments with fractional Gaussian noises. Part 2, rescaled ranges and spectra *Water Resources Res.* **5** 242–59 (Ch 4)

——1969c Computer experiments with fractional Gaussian noises. Part 3, mathematical appendix *Water Resources Res.* **5** 260–7 (Ch 4)

——1969d Some long-run properties of geophysical records *Water Resources Res.* **5** 321–40 (Ch 4)

Manneville P, Boccara N, Vichniac G Y and Bidaux R (eds) 1989 Cellular automata and modeling of complex physical systems *Proc. Winter School (Les Houches, 1989)* (Berlin: Springer) (Ch 6)

Marek M and Schreiber I 1991 Chaotic behaviour of deterministic dissipative systems (Cambridge: Cambridge University Press) (Ch 6)

Massopust P R 1994 *Fractal Functions, Fractal Surfaces, and Wavelets* (San Diego, CA: Academic) (Ch 3)

Matthews P C, Mirollo R E and Strogatz S H 1991 Dynamics of a large system of coupled nonlinear oscillators *Physica* D **52** 293–331 (Ch 6)

Mattila P 1995 *Geometry of Sets and Measures in Euclidean Spaces: Fractals and Rectifiability* (Cambridge: Cambridge University Press) (Ch 3)

May R M 1974 Biological populations with nonoverlapping generations: stable points, stable cycles, and chaos *Science* **186** 645–7 (Ch 5)

——1976 Simple mathematics with very complicated dynamics *Nature* **261** 459–67 (Ch 5)

Mayer-Kress G (ed) 1986 Dimensions and entropies in chaotic systems: quantification of complex behaviour *Proc. Int. Workshop (Pecos River Ranch, 1985)* (Berlin: Springer) (Ch 7)

——1987 Application of dimension algorithms to experimental chaos *Directions in Chaos: Volume 1* ed H Bai-Lin (Singapore: World Scientific) (Ch 7)

Mayer-Kress G, Bargar R and Choi I 1992 Musical structures in data from chaotic attractors *University of Illinois Technical Report UIUC-BI-CCSR-92-14* (Ch 5)

Mayer-Kress G, Choi I, Weber N, Bargar R and Hübler A 1993 Musical signals from Chua's circuit *University of Illinois Technical Report UIUC-BI-CCSR-93-04* (Ch 6)

McCauley J L 1993 *Chaos, Dynamics, and Fractals: an Algorithmic Approach to Deterministic Chaos* (Cambridge: Cambridge University Press) (Ch 6)

Meakin P 1983 Diffusion-controlled cluster formation in 2–6-dimensional space *Phys. Rev.* A **27** 1495–507 (Ch 4)

——1991 Models for material failure and deformation *Science* **252** 226–34 (Ch 4)

Megaridis C M and Dobbins R A 1990 Morphological description of flame-generated materials *Combustion Sci. Technol.* **71** 95–109 (Ch 4)

Mehra V and Ramaswamy R 1996 Maximal Lyapunov exponent at crises *Phys. Rev.* E **53** 3420–4 (Ch 7)

Meneveau C and Sreenivasan K R 1987 Simple multifractal cascade model for fully developed turbulence *Phys. Rev. Lett.* **59** 1424–7 (Ch 6)

——1991 The multifractal nature of turbulent energy dissipation *J. Fluid Mech.* **224** 429–84 (Ch 6, Ch 7)

Mercader I, Massaguer J M and Net M 1990 Hysteresis in a transition to chaos *Phys. Lett.* **149A** 195–9 (Ch 6)

Metropolis N, Stein M L and Stein P R 1973 On finite limit sets for transformations on the unit interval *J. Combinatorial Theory* **15** 25–44 (Ch 5)

Metzler W 1994 The 'mystery' of the quadratic Mandelbrot set *Am. J. Phys.* **62** 813–4 (Ch 5)

Middleton G V 1990 Non-linear dynamics and chaos: potential applications in the earth sciences *Geosci. Canada* **17** 3–11 (Ch 6)

Miles J 1983 Strange attractors in fluid dynamics *Adv. Appl. Mech.* **24** 189–214 (Ch 6)

Miller G T 1993 A tutorial on the visualization of forward orbits associated with Seigel disks in the quadratic Julia sets *Comput. Graph.* **17** 321–4 (Ch 5)

Milovanov A V and Zelenyi L M 1993 Applications of fractal geometry to dynamical evolution of sunspots *Phys. Fluids* B **5** 2609–15 (Ch 3)

Min S W, Lim S P and Lee K S 1995 A fractal surface and its measurement by computer simulation *Opt. Laser Technol.* **27** 331–3 (Ch 3)

Mira C 1987 *Chaotic Dynamics: From the One-Dimensional Endomorphism to the Two-Dimensional Diffeomorphism* (Singapore: World Scientific) (Ch 6)

Mirasso C R, Colet P and García-Fernández P 1996 Synchronization of chaotic semiconductor lasers: application to encoded communications *IEEE Photon. Technol. Lett.* **8** 299–301 (Ch 6)

Mitschke F, Möller M and Lange W 1988 Measuring filtered chaotic signals *Phys. Rev.* A **37** 4518–21 (Ch 7)

Moon F C 1980 Experiments on chaotic motions of a forced nonlinear oscillator: strange attractors *Trans. ASME J. Appl. Mech.* **47** 638–44 (Ch 6)

——1987 *Chaotic Vibrations: an Introduction for Applied Scientists and Engineers* (New York: Wiley) (Ch 6)

Moon F C and Holmes P J 1979 A magnetoelastic strange attractor *J. Sound Vib.* **65** 275–96 (Ch 6)

Moon F C and Li G-X 1985a Fractal basin boundaries and homoclinic orbits for periodic motion in a two-well potential *Phys. Rev. Lett.* **55** 1439–42 (Ch 6)

——1985b The fractal dimension of the two-well potential strange attractor *Physica* D **17** 99–108 (Ch 6, Ch 7)

Moore C A and Krepfl M 1991 Using fractals to model soil fabric *Geotechnique* **41** 123–34 (Ch 3)

Moorthy R I K, Kakodkar A and Srirangarajan H R 1996 The significance of higher modes for evolution of chaos in structural mechanics systems *J. Sound Vib.* **198** 267–77 (Ch 6)

Moss F and McClintock P V E 1989 *Noise in Nonlinear Dynamical Systems. Volume 3: Experiments and Simulations* (Cambridge: Cambridge University Press) (Ch 7)

Mu Z Q and Lung C W 1988 Studies on the fractal dimension and fracture toughness of steel *J. Phys. D: Appl. Phys.* **21** 848–50 (Ch 3, Ch 4)

Müller P C, Bajkowski J and Söffker D 1994 Chaotic motions and fault detection in a cracked rotor *Nonlinear Dynamics* **5** 233–54 (Ch 6)

Müller S C, Coullet P and Walgraef D 1994 From oscillations to excitability: a case study in spatially extended systems *Chaos* **4** 439–42 (Ch 6, Ch 7) (This is the introductory paper in a special focus issue of the journal on spatially extended systems.)

Mullin T (ed) 1993 *The Nature of Chaos* (Oxford: Clarendon) (Ch 6)

Mullin T and Price T J 1989 An experimental observation of chaos arising from the interaction of steady and time-dependent flows *Nature* **340** 294–6 (Ch 6)

Nakagawa M 1993 A critical exponent method to evaluate fractal dimensions of self-affine data *J. Phys. Soc. Japan* **62** 4233–9 (Ch 4)

Nakamura K 1993 *Quantum Chaos* (Cambridge: Cambridge University Press) (Ch 6)

Navé P 1994 A literary ancestor of Lorenz's butterfly *Phys. Today* December 85 (Ch 6)

Néda Z, Bakó B and Rees E 1996 The dripping faucet revisited *Chaos* **6** 59–62 (Ch 6)

Neilson R D and Gonsalves D H 1992 Chaotic motion of a rotor system with a bearing clearance *Applications of Chaos and Fractals—Int. State-of-the-Art Seminar (London, 1992)* (Ch 6)

Nettel S 1995 *Wave Physics: Oscillations—Solitons— Chaos* (Berlin: Springer) (Ch 6)

Newland D E 1993 *An Introduction to Random Vibrations, Spectral and Wavelet Analysis* (Harlow: Longman) (Ch 7)

Nezlin M V 1994 Some remarks on coherent structures out of chaos in planetary atmospheres and oceans *Chaos* **4** 109–11 (This is the introductory paper in a special focus issue of the journal on 'large long-lived coherent structures out of chaos in planetary atmospheres and oceans') (Ch 6)

Nicolis J S 1991 *Chaos and Information Processing: a Heuristic Outline* (Singapore: World Scientific) (Ch 6)

Niemeyer L, Pietronero L and Wiesmann H J 1984 Fractal dimension of dielectric breakdown *Phys. Rev. Lett.* **52** 1033–6 (Ch 4)

Niklasson G A 1993 Adsorption of fractal structures: applications to cement materials *Cement Concrete Res.* **23** 1153 8 (Ch 3)

Nikora V I, Sapozhnikov V B and Noever D A 1993 Fractal geometry of individual river channels and its computer simulation *Water Resources Res.* **29** 3561–8 (Ch 3)

Nittmann J, Daccord G and Stanley H E 1985 Fractal growth of viscous fingers: quantitative characterization of a fluid instability phenomenon *Nature* **314** 141–4 (Ch 4)

Noack B R and Eckelmann H 1992 On chaos in wakes *Physica* D **56** 151–64 (Ch 6)

Nonnenmacher T F, Losa G A and Weibel E R (eds) 1994 *Fractals in Biology and Medicine* (Basel: Birkhäuser) (Ch 3)

Novikov E A 1991 Chaotic vortex–body interaction *Phys. Lett.* **152A** 393–6 (Ch 6)

Ogorzalek M (ed) 1995 Nonlinear dynamics of electronic systems *Int. J. Electron.* **79** (Ch 6)

Okubo A 1971 Oceanic diffusion diagrams *Deep Sea Res.* **18** 789–802 (Ch 4)

Olsen L F and Degn H 1977 Chaos in an enzyme reaction *Nature* **267** 177–8 (Ch 6)

——1985 Chaos in biological systems *Q. Rev. Biophys.* **18** 165–225 (Ch 6)

Orbach R 1986 Dynamics of fractal networks *Science* **231** 814–9 (Ch 4)

Osborne A R, Kirwan A D, Provenzale A and Bergamasco L 1989 Fractal drifter trajectories in the Kuroshio extension *Tellus* **41** 416–35 (Ch 3, Ch 4)

Osborne A R and Provenzale A 1989 Finite correlation dimension for stochastic systems with power-law spectra *Physica* D **35** 357–81 (Ch 4)

Ott E 1993 *Chaos in Dynamical Systems* (Cambridge: Cambridge University Press) (Ch 6, Ch 7)

——1994 Recent developments in chaotic dynamics *IEEE Trans. Plasma Sci.* **22** 43–6 (Ch 7)

Ott E, Grebogi C and Yorke J A 1990 Controlling chaos *Phys. Rev. Lett.* **64** 1196–9 (Ch 7)

Ott E, Sauer T and Yorke J A 1994 *Coping with Chaos: Analysis of Chaotic Data and the Exploitation of Chaotic Systems* (New York: Wiley) (Ch 7)

Ott E and Tél T 1993 Chaotic scattering: an introduction *Chaos* **3** 417–26 (This is the introductory paper in a special focus issue of the journal on 'chaotic scattering'.) (Ch 6)

Ottino J M 1989 *The Kinematics of Mixing: Stretching, Chaos, and Transport* (Cambridge: Cambridge University Press) (Ch 6)

Ozorio de Almeida A M 1988 *Hamiltonian Systems; Chaos and Quantization* (Cambridge: Cambridge University Press) (Ch 6)

Packard N H, Crutchfield J P, Farmer J D and Shaw R S 1980 Geometry from a time series *Phys. Rev. Lett.* **45** 712–6 (Ch 7)

Pahl-Wostl C 1995 *The Dynamic Nature of Ecosystems: Chaos and Order Entwined* (Chichester: Wiley) (Ch 6)

Paidoussis M P, Li G X and Moon F C 1989 Chaotic oscillations of the autonomous system of a constrained pipe conveying fluid *J. Sound Vib.* **135** 1–19 (Ch 6)

Paidoussis M P and Moon F C 1988 Nonlinear and chaotic fluidelastic vibrations of a flexible pipe conveying fluid *J. Fluids Struct.* **2** 567–91 (Ch 6)

Paladin G and Vulpiani A 1987 Anomalous scaling laws in multifractal objects *Phys. Rep.* **156** 147–225 (Ch 7)

Paladino O 1994 Transition to chaos in continuous processes: applications to wastewater treatment reactors *Environmetrics* **5** 57–70 (Ch 6)

Palis J and Takens F 1993 *Hyperbolicity and Sensitive Chaotic Dynamics at Homoclinic Bifurcations* (Cambridge: Cambridge University Press) (Ch 6)

Pande C S, Richards L R and Smith S 1987 Fractal characteristics of fractured surfaces *J. Mater. Sci. Lett.* **6** 295–7 (Ch 3)

Pando C L, Acosta G A L, Meucci R and Ciofini M 1995 Highly dissipative Hénon map behaviour in the four-level model of the CO_2 laser with modulated losses *Phys. Lett.* **199A** 191–8 (Ch 5)

Parker D and Stacey R 1994 *Chaos, Management and Economics: The Implications of Non-Linear Thinking* (London: Institute of Economic Affairs) (Ch 6)

Parlitz U 1992 Identification of true and spurious Lyapunov exponents from time series *Int. J. Bifurcation Chaos* **2** 155–65 (Ch 7)

Peebles P J E 1989 The fractal galaxy distribution *Physica* D **38** 273–8 (Ch 3)

Peitgen H-O, Jürgens H and Saupe D 1992a *Chaos and Fractals: New Frontiers of Science* (New York: Springer) (Ch 2)

Peitgen H-O, Jürgens H, Saupe D, Maletsky E M, Perciante T H and Yunker L E 1991 *Fractals for the Classroom: Strategic Activities Volume One* (New York: Springer) (Ch 2)

——1992b *Fractals for the Classroom: Strategic Activities Volume Two* (New York: Springer) (Ch 5)

Peitgen H-O, Jurgens H, Saupe D and Zhalten C (eds) 1990 *Fractals: An Animated Discussion* (A Video) (New York: Freeman) (Ch 1, Ch 5)

Peitgen H-O and Richter P H 1986 *The Beauty of Fractals: Images of Complex Dynamical Systems* (Berlin: Springer) (Ch 5)

Peitgen H-O and Saupe D (eds) 1988 *The Science of Fractal Images* (New York: Springer) (Ch 3, Ch 4)

Peleg M 1993 Fractals and foods *Crit. Rev. Food Sci. Nutr.* **33** 149–65 (Ch 3)

Peng C-K, Buldyrev S V, Goldberger A L, Havlin S, Sciortino F, Simons M and Stanley H E 1992 Long-range correlations in nucleotide sequences *Nature* **356** 168–70 (Ch 4)

Peng J H, Ding E J, Ding M and Yang W 1996 Synchronizing hyperchaos with a scalar transmitted signal *Phys. Rev. Lett.* **76** 904–7 (Ch 6)

Pesaran M H and Potter S M 1993 *Nonlinear Dynamics, Chaos and Econometrics* (Chichester: Wiley) (Ch 6)

Pezeshki C and Dowell E H 1987 An examination of initial condition maps for the sinusoidally excited buckled beam modelled by the Duffing's equation *J. Sound Vib.* **117** 219–32 (Ch 6)

Phatak S C and Rao S S 1995 Logistic map: a possible random-number generator *Phys. Rev.* E **51** 3670–8 (Ch 5)

Philip A G D, Frame M and Robucci A 1994 Warped midgets in the Mandelbrot set *Comput. Graph.* **18** 239–48 (Ch 5)

Philip K W 1992 Field lines in the Mandelbrot set *Comput. Graph.* **16** 443–7 (Ch 5)

Pickover C A 1992 Chaos and graphics introduction by the associate editor *Comput. Graph.* **16** 3–7 (Ch 1)

——1994 Automatic parallel generation of Aeolian fractals on the IBM Power Visualization System *Comput. Graph.* **18** 407–16 (Ch 5)

——(ed) 1995 *The Pattern Book: Fractals Art and Nature* (Singapore: World Scientific) (Ch 5)

Pierson D and Moss F 1995 Detecting periodic unstable points in noisy chaotic and limit cycle attractors with applications to biology *Phys. Rev. Lett.* **75** 2124–7 (Ch 7)

Pietronero L and Tossatti E (eds) 1986 *Fractals in Physics* (Amsterdam: North Holland) (Ch 3)

Pires A, Landau D P and Herrman H (eds) 1990 *Workshop on Computational Physics and Cellular Automata (Ouro Preto, 1989)* (Singapore: World Scientific) (Ch 6)

Plaschko P, Berger E and Brod K 1993 The transition of flow-induced cylinder vibrations to chaos *Nonlinear Dynam.* **4** 251–268 (Ch 6)

Poddar B, Moon F C and Mukherjee S 1988 Chaotic motion of an elastic-plastic beam *ASME J. Appl. Mech.* **55** 185–9 (Ch 6)

Pomeau Y and Manneville P 1980 Intermittent transition to turbulence in dissipative dynamical systems *Commun. Math. Phys.* **74** 189–97 (Ch 6)

Poon D C, McCormack M and Thimm H F 1993 The application of fractal geostatistics to oil and gas property evaluation and reserve estimates *J. Can. Petroleum Technol.* **32** 24–7 (Ch 3)

Pottinger D, Todd S, Rodrigues I, Mullin T and Skeldon A 1992 Phase portraits for parametrically excited pendula: an exercise in multidimensional data visualisation *Comput. Graph.* **16** 331–7 (Ch 7)

Press W H, Teukolsky S A, Vetterling W T and Flannery B.P. 1992 *Numerical Recipes in FORTRAN: the Art of Scientific Computing* 2nd edn (Cambridge: Cambridge University Press) (Ch 6, Ch 7)

Pressing J 1988 Nonlinear maps as generators of musical design *Comput. Music J.* **12** 35–46 (Ch 5)

Procaccia I 1988 Complex or just complicated? *Nature* **333** 498–9 (Ch 7)

Pumar M A 1996 Zooming of terrain imagery using fractal-based interpolation *Comput. Graph.* **20** 171–6 (Ch 4)

Qammar H and Mossayebi F 1996 Fractal basins in the control of the logistic equation *Comput. Graph.* **20** 589–96 (Ch 5)

Rahman Z and Burton T D 1986 Large amplitude primary and superharmonic resonances in the Duffing oscillator *J. Sound Vib.* **110** 363–80 (Ch 6)

Rajasekar S and Lakshmanan M 1993 Algorithms for controlling chaotic motion: application for the BVP oscillator *Physica* D **67** 282–300 (Ch 7)

Rambaldi S and Pinazza O 1994 An accurate fractional Brownian motion generator *Physica* A **208** 21–30 (Ch 4)

Ramirez R W 1985 *The FFT: Fundamentals and Concepts* (Engelwood Cliffs, NJ: Prentice-Hall) (Ch 7)

Rapaport D C 1985 The fractal nature of molecular trajectories in fluids *J. Stat. Phys.* **40** 751–8 (Ch 4)

Ravindra B and Mallik A K 1994 Stability analysis of a non-linearly damped Duffing oscillator *J. Sound Vib.* **171** 708–16 (Ch 6)

Reed I S, Lee P C and Truong T K 1995 Spectral representation of fractional Brownian motion in *n* dimensions and its properties *IEEE Trans. Information Theory* **41** 1439–51 (Ch 4)

Reis F D A A 1995a Finite-size scaling for random walks on fractals *J. Phys. A: Math. Gen.* **28** 6277–87 (Ch 4)

——1995b Non-universal behaviour of self attracting walks *J. Phys. A: Math. Gen.* **28** 3851–62 (Ch 4)

Reiter C A 1994 Sierpinski fractals and GCD's *Comput. Graph.* **18** 885–91 (Ch 2)

Renshaw E 1994 Chaos in biometry *IMA J. Math. Appl. Med. Biol.* **11** 17–44 (Ch 6)

Reynolds R R, Virgin L N and Dowell E H 1993 High-dimensional chaos can lead to weak turbulence *Nonlinear Dynam.* **4** 531–46 (Ch 6)

Richardson L F 1926 Atmospheric diffusion shown on a distance-neighbour graph *Proc. R. Soc.* A **110** 709–37 (Ch 4)

——1961 The problem with contiguity: an appendix to statistics of deadly quarrels *Gen. Syst. Yearbook* **6** 139–87 (Ch 3)

Röder H, Hahn E, Brune H, Bucher J-P and Kern K 1993 Building one- and two-dimensional nanostructures by diffusion-controlled aggregation at surfaces *Nature* **366** 141–3 (Ch 4)

Roman H E and Alemany P A 1994 Continuous-time random walks and the fractional diffusion equation *J. Phys. A: Math. Gen.* **27** 3407–10 (Ch 4)

Romeiras F J, Bondeson A, Ott E, Antonsen T M and Grebogi C 1987 Quasiperiodically forced dynamcial systems with strange nonchaotic attractors *Physica* D **26** 277–94 (Ch 7)

Romera M, Pastor G and Montoya F 1996 Graphic tools to analyse one-dimensional quadratic maps *Comput. Graph.* **20** 333–9 (Ch 5)

Rosenblum M G, Pikovsky A S and Kurths J 1996 Phase synchronization of chaotic oscillators *Phys. Rev. Lett.* **76** 1804–7 (Ch 6)

Rosenstein M T and Collins J J 1994 Visualizing the effects of filtering chaotic signals *Comput. Graph.* **18** 587–92 (Ch 7)

Rosenstein M T, Collins J J and De Luca C J 1993 A practical method for calculating largest Lyapunov exponents from small data sets *Physica* D **65** 117–34 (Ch 7)

——1994 Reconstruction expansion as a geometry-based framework for choosing proper delay times *Physica* D **73** 82–98 (Ch 7)

Rössler O E 1976 An equation for continuous chaos *Phys. Lett.* **57A** 397–8 (Ch 6)

——1979 An equation for hyperchaos *Phys. Lett.* **71A** 155–7 (Ch 6)

Roux J-C 1983 Experimental studies of bifurcations leading to chaos in the Belousof–Zhabotinsky reaction *Physica* D **7** 57–68 (Ch 6)

Roux J-C, Rossi A, Bachelart S and Vidal C 1980 Representation of a strange attractor from an experimental study of chemical turbulence *Phys. Lett.* **77A** 391–3 (Ch 7)

Roux J-C, Simoyi R H and Swinney H L 1983 Observation of a strange attractor *Physica* D **8** 257–66 (Ch 6)

Roux J-C and Swinney H L 1981 Topology of chaos in a chemical reaction *Nonlinear Phenomena in Chemical Dynamics: Proc. Int. Conf. (Bordeaux, 1981)* ed C Vidal and A Pacault (Berlin: Springer) pp 38–43 (Ch 6)

Roy A E and Steves B A (eds) 1995 *From Newton to Chaos: Modern Techniques for Understanding and Coping with Chaos in N-Body Dynamical Systems* (New York: Plenum) (Ch 6)

Roy R, Murphy T W, Maier T D and Gills Z 1992 Dynamical control of a chaotic laser: experimental stabilization of a globally coupled system *Phys. Rev. Lett.* **68** 1259–62 (Ch 7)

Ruelle D 1980 Strange attractors *Math. Intel.* **2** 126–37 (Ch 6)

——1983a Turbulent dynamical systems *Proc. Int. Cong. Math. (Warsaw, 1983)* 271–86 (Ch 6)

——1983b Five turbulent problems *Physica* D **7** 40–4 (Ch 6)

——1989 *Chaotic Evolution and Strange Attractors* (Cambridge: Cambridge University Press) (Ch 6)

——1990 Deterministic chaos: the science and the fiction *Proc. R. Soc.* A **427** 241–8 (Ch 7)

——1993 *Chance and Chaos* (London: Penguin) (Ch 1)

Ruelle D and Takens F 1971 On the nature of turbulence *Commun. Math. Phys.* **20** 167–92 (Ch 6)

Rulkov N F and Shushchik M M 1996 Experimental observation of synchronised chaos with frequency ratio 1:2 *Phys. Lett.* **214A** 145–50 (Ch 6)

Russell D A, Hanson J D and Ott E 1980 Dimension of strange attractors *Phys. Rev. Lett.* **45** 1175–8 (Ch 7)

Rys F S and Waldvogel A 1986 Fractal shape of hail clouds *Phys. Rev. Lett.* **56** 784–7 (Ch 3)

Sagdeev R Z, Usikov D A and Zaslavsky G M 1988 *Nonlinear Physics: From the Pendulum to Turbulence and Chaos* (Chur: Harwood) (Ch 6)

Sahimi M 1993 Fractal and superdiffusive transport and hydrodynamic dispersion in heterogeneous porous media *Transport Porous Media* **13** 3–40 (Ch 4)

Saltzman B 1962 Finite amplitude free convection as an initial value problem—I *J. Atmospher. Sci.* **19** 329–41 (Ch 6)

Sanderson B G and Booth D A 1991 The fractal dimension of drifter trajectories and estimates of horizontal eddy-diffusivity *Tellus* A **43** 334–49 (Ch 3, Ch 4)

Sano M and Sawada Y 1985 Measurement of the Lyapunov spectrum from a chaotic time series *Phys. Rev. Lett.* **55** 1082–5 (Ch 7)

Sarathy R and Sachdev P L 1994 On the gluing and ungluing of strange attractors in the study of the Lorenz system *Phys. Lett.* **191A** 238–44 (Ch 6)

Sarkar S (ed) 1986 *Nonlinear Phenomena and Chaos (Malvern Physics Series)* (Bristol: Hilger) (Ch 6)

Sauer T, Yorke J A and Casdagli M 1991 Embedology *J. Stat. Phys.* **65** 579–616 (Ch 7)

Scheck F 1990 *Mechanics: From Newton's Laws to Deterministic Chaos* (Berlin: Springer) (Ch 6)

Schmid G B and Dünki R M 1996 Indications of nonlinearity, intraindividual specificity and stability of human EEG: the unfolding dimension *Physica* D **93** 165–90 (Ch 6)

Schroeder M 1991 *Fractals, Chaos, Power Laws—Minutes from an Infinite Paradise* (New York: Freeman) (Ch 3)

Schuster H G 1988 *Deterministic Chaos: an Introduction* 2nd revised edn (Weinheim: VCH) (Ch 6)

Scott S K 1991 *Chemical Chaos* (Oxford: Clarendon) (Ch 6)

——1994 *Oscillations, Waves and Chaos in Chemical Kinetics* (Oxford: Oxford University Press) (Ch 6)

Seimenis J (ed) 1994 *Hamiltonian Mechanics: Integrability and Chaotic Behaviour* (New York: Plenum) (Ch 6)

Seydel R 1988 *From Equilibrium to Chaos: Practical Bifurcation and Stability Analysis* (New York: Elsevier) (Ch 6)

Shaw R 1984 *The Dripping Faucet as a Model Chaotic System* (Santa Cruz, CA: Aerial) (Ch 6)

Sherard P K 1993 Julia sets and quasi-stable orbits in the complex plane *Comput. Graph.* **17** 175–84 (Ch 5)

Shishikura M 1994 The boundary of the Mandelbrot set has Hausdorff dimension two *Asterisque* No 222 389–405 (Ch 5)

Shlesinger M F 1989 Levy flights: variations on a theme *Physica* D **38** 304–9 (Ch 4)

Singer J, Wang Y-Z and Bau H H 1991 Controlling a chaotic system *Phys. Rev. Lett.* **66** 1123–5 (Ch 7)

Smallwood G J, Gülder Ö L, Snelling D R, Deschamps B M and Gökalp I 1995 Characterization of flame front surfaces in turbulent premixed methane/air combustion *Combustion Flame* **101** 461–70 (Ch 3)

Smith L A 1988 Intrinsic limits on dimension calculations *Phys. Lett.* **133A** 283–8 (Ch 7)

Soliman M S and Thompson J M T 1991 Transient and steady state analysis of capsize phenomena *Appl. Ocean Res.* **13** 82–92 (ch 6)

Sparrow C 1982 *The Lorenz Equations: Bifurcations, Chaos, and Strange Attractors* (New York: Springer) (Ch 6)

Sprott J C 1993 Automatic generation of strange attractors *Comput. Graph.* **17** 325–32 (Ch 5)

Sprott J C and Pickover C A 1995 Automatic generation of general quadratic map basins *Comput. Graph.* **19** 309–13 (Ch 5)

Sreenivasan K R 1991 Fractals and multifractals in fluid turbulence *Ann. Rev. Fluid Mech.* **23** 539–600 (Ch 3)

Sreenivasan K R and Prasad R R 1989 New results on the fractal and multifractal structure of the large Schmidt number passive scalars in fully turbulent flows *Physica* D **38** 322–9 (Ch 6)

Sreenivasan K R and Ramshankar R 1986 Transition intermittency in open flows, and intermittency routes to chaos *Physica* D **23** 246–58 (Ch 6)

Sri Namachchivaya N, Van Roessel H J and Talwar S 1994 Maximal Lyapunov exponent and almost-sure stability for coupled two-degree-of-freedom stochastic systems *Trans. ASME: J. Appl. Mech.* **61** 446–52 (Ch 7)

Stanley H E 1989 Learning concepts of fractals and probability by 'doing science' *Physica* D **38** 330–40 (Ch 4)

Stanley H E and Ostrowsky N 1986 *On Growth and Form: Fractal and Non-Fractal Patterns in Physics* (Dordrecht: Nijhoff) (Ch 4)

Steeb W-H and Louw J A 1986 *Chaos and Quantum Chaos* (Singapore: World Scientific) (Ch 6)

Stewart I 1989 *Does God Play Dice?: the Mathematics of Chaos* (Oxford: Blackwell) (Ch 1)

Strogatz S H 1994 *Nonlinear Dynamics and Chaos* (Reading, MA: Addison-Wesley) (Ch 6)

Stuart A M and Humphries A R 1996 *Dynamical Systems and Numerical Analysis* (Cambridge: Cambridge University Press) (Ch 6)

Sugihara G and May R M 1990 Nonlinear forcasting as a way of distinguishing chaos from measurement error in time series *Nature* **344** 734–40 (Ch 7)

Sussman G J and Wisdom J 1988 Numerical evidence that the motion of Pluto is chaotic *Massachusetts Institute of Technology Artificial Intelligence Laboratory, AI Memo* **1039** (Ch 6)

——1992 Chaotic evolution of the solar system *Massachusetts Institute of Technology Artificial Intelligence Laboratory, AI Memo* **1359** (Ch 6)

Swinney H L and Gollub J P (eds) 1981 *Hydrodynamic Instabilities and the Transition to Turbulence* (Berlin: Springer) (Ch 6)

Sykes R I and Gabruk R S 1994 Fractal representation of turbulent dispersing flumes *J. Appl. Meteorol.* **33** 721–32 (Ch 3)

Szemplinska-Stupnicka W 1987 Secondary resonances and approximate models of routes to chaotic motion in non-linear oscillators *J. Sound Vib.* **113** 155–72 (Ch 6)

——1994 A discussion of an analytical method of controlling chaos in Duffing's oscillator *J. Sound Vib.* **178** 276–84 (Ch 6)

Szpiro G G 1993 Measuring dynamical noise in dynamical systems *Physica* D **65** 289–99 (Ch 7)

Tabeling P, Cardoso O and Perrin B 1990 Chaos in a linear array of vortices *J. Fluid Mech.* **213** 511–30 (Ch 6)

Takens F 1981 *Detecting Strange Attractors in Turbulence (Lecture Notes in Mathematics 898)* ed D A Rand and L S Young (Berlin: Springer) pp 366–81 (Ch 7)

Talibuddin S and Runt J P 1994 Reliability test of popular fractal techniques applied to small two-dimensional self-affine data sets *J. Appl. Phys.* **76** 5070–8 (Ch 4)

Tan B L and Chia T T 1995 Unusual period doublings in the linear-logistic map *Phys. Rev.* E **52** 6885–8 (Ch 5)

Tatsumi T (ed) 1984 *Turbulence and Chaotic Phenomena in Fluids* (Amsterdam: Elsevier) (Ch 6)

Testa J, Pérez J and Jeffries C 1982 Evidence for universal chaotic behaviour of a driven nonlinear oscillator *Phys. Rev. Lett.* **48** 714–8 (Ch 6)

Theiler J 1986 Spurious dimension from correlation algorithms applied to limited time-series data *Phys. Rev.* A **34** 2427–32 (Ch 7)

Theiler J, Eubank S, Longtin A, Galdrikian B and Farmer J D 1992 Testing for nonlinearity in time series: the method of surrogate data *Physica* D **58** 77–95 (Ch 7)

Thiran P and Setti G 1995 Information storage and retrieval in an associative memory based on one-dimensional maps *Int. J. Electron.* **79** 815–21 (Ch 5) (Note that this paper is part of a special issue of the journal concerned with the nonlinear dynamics of electronic circuits.)

Thompson J M T and Stewart H B 1986 *Nonlinear Dynamics and Chaos* (Chichester: Wiley) (Ch 6)

Tong H and Smith R L 1992 Royal Statistical Society Meeting on Chaos: Introduction *J. R. Stat. Soc.* B **54** 301–2 (This is the introductory paper in a series of papers in this volume reporting on 'The Royal Statistical Society meeting on chaos' on October 6 1991) (Ch 6)

Tricot C 1995 *Curves and Fractal Dimension* (New York: Springer) (Ch 3)

Tritton D J 1986 Ordered and chaotic motion of a forced spherical pendulum *Eur. J. Phys.* **7** 162–9 (Ch 6)

——1988 *Physical Fluid Dynamics* 2nd edn (Oxford: Clarendon) (Ch 6)

Tritton D J and Egdell C 1993 Chaotic bubbling *Phys. Fluids* A **5** 503–5 (Ch 6)

Tseng W-Y and Dugundji J 1971 Nonlinear vibrations of a buckled beam under harmonic excitation *Trans. ASME J. Appl. Mech.* **38** 467–76 (Ch 6)

Tsonis A A 1992 *Chaos: From Theory to Applications* (New York: Plenum) (Ch 6)

Tufillaro N B, Abbott T and Reilly J 1992 *An Experimental Approach to Nonlinear Dynamics and Chaos* (Redwood City, CA: Addison-Wesley) (Ch 7)

Turcotte D L 1992 *Fractals and Chaos in Geology and Geophysics* (Cambridge: Cambridge University Press) (Ch 3, Ch 6)

Turner J S, Roux J-C, McCormick W D and Swinney H L 1981 Alternating periodic and chaotic regimes in a chemical reaction—experiment and theory *Phys. Lett.* **85A** 9–12 (Ch 6)

Ueda Y 1979 Randomly transitional phenomena in the system governed by Duffing's equation *J. Stat. Phys.* **20** 181–96 (Ch 6)

——1980a Steady motions exhibited by Duffing's equation: a picture book of regular and chaotic motions *New Approaches to Nonlinear Problems in Dynamics* ed P J Holmes (Philadelphia, PA: SIAM) (Ch 6).

——1980b Explosion of strange attractors exhibited by Duffing's equation *Ann. NY Acad. Sci.* **357** 422–34 (Ch 6)

Vandewalle N and Ausloos M 1995 The robustness of self-organized criticality against extinctions in a tree-like model of evolution *Europhys. Lett.* **32** 613–8 (Ch 3)

Vaneck T 1994 The hyper-volume method: finding the boundaries between regular and chaotic phenomena in the forced Duffing oscillator *J. Sound Vib.* **175** 570–6 (Ch 6)

Velarde M G and Antoranz J C 1981 Strange attractor (optical turbulence) in a model problem for the laser with saturable absorber and the two-component Bénard convection *Prog. Theor. Phys.* **66** 717–20 (Ch 6)

Vicsek T 1989 *Fractal Growth Phenomena* (Singapore: World Scientific) (Ch 4)

Vieux B E and Farajalla N S 1994 Capturing the essential spatial variability in distributed hydrological modelling: hydraulic roughness *Hydrolog. Processes* **8** 221–36 (Ch 3)

Villermaux J 1993 Future challenges for basic research in chemical engineering *Chem. Eng. Sci.* **48** 2525–35 (Ch 6)

Voss R F 1985 Random fractal forgeries *Fundamental Algorithms for Computer Graphics (NATO ASI Series)* ed R A Earnshaw (Berlin: Springer) pp 805–35 (Ch 4)

——1989 Random fractals: self affinity in music, mountains, and clouds *Physica* D **38** 362–71 (Ch 4)

——1992 Multifractals and the local connected fractal dimension: classification of early Chinese landscape drawings *Applications of Chaos and Fractals—Int. State-of-the-Art Seminar (London, 1992)* (Ch 3)

——1993 Reply to Buldyrev S V, Goldberger A L, Havlin S, Peng C-K, Simons M, Sciortino F and Stanley H E 1993 Long-range power-law correlations in DNA *Phys. Rev. Lett.* **71** 1776, 1777 (Ch 4)

Waldner F, Barberis D R and Yamazaki H 1985 Route to chaos by irregular periods: simulations of parallel pumping in ferromagnets *Phys. Rev.* A **31** 420–31 (Ch 6)

Walter D 1994 Computer art from Newton's, secant, and Richardson's method *Comput. Graph.* **18** 127–33 (Ch 5)

Wang F and Cohen E G D 1996 Diffusion on random lattices *J. Stat. Phys.* **84** 233–61 (Ch 4)

Wang K G 1994 Long-range correlation effects, generalized Brownian motion and anomolous diffusion *J. Phys. A: Math. Gen.* **27** 3655–61 (Ch 4)

Wang X-B, Li J-X, Jiang Q, Zhang Z-H and Tian D-C 1994 Effects of fractons in superconductors with fractal structure *Phys. Rev.* B **49** 9778–81 (Ch 3)

West B J 1990 *Fractal Physiology and Chaos in Medicine: Studies in Nonlinear Phenomena in Life Sciences—Vol 1* (Singapore: World Scientific) (Ch 3, Ch 6)

——1995 Fractal statistics in biology *Physica* D **86** 12–8 (Ch 4)

West B J and Deering W 1994 Fractal physiology for physicists: Lévy statistics *Phys. Rep.* **246** 1–100 (Ch 4)

West B J and Goldberger A L 1987 Physiology in fractal dimensions *Am. Sci.* **75** 354–65 (Ch 3)

White M T and Tongue B H 1995 Application of interpolated cell mapping to an analysis of the Lorenz equations *J. Sound Vib.* **188** 209–26 (Ch 6)

Whitney H 1936 Differentiable manifolds *Ann. Math.* **37** 645–80 (Ch 7)

Wicks K R 1991 *Fractals and Hyperspaces (Lecture Notes in Mathematics 1492)* (Berlin: Springer) (Ch 2, Ch 3)

Wiggins S 1987 Chaos in the quasiperiodically forced Duffing oscillator *Phys. Lett.* **124A** 138–42 (Ch 6)

——1988 *Global Bifurcations and Chaos: Analytical Methods* (New York: Springer) (Ch 6)

——1990 *Introduction to Applied Nonlinear Dynamical Systems and Chaos* (New York: Springer) (Ch 6)

Williams-Stuber K and Gharib M 1990 Transition from order to chaos in the wake of an airfoil *J. Fluid Mech.* **213** 29–57 (Ch 6)

Witten T A and Sander L M 1981 Diffusion limited aggregation, a kinetic critical phenomenon *Phys. Rev. Lett.* **47** 1400–3 (Ch 4)

Wolf A, Swift J B, Swinney H L and Vastano J A 1985 Determining Lyapunov exponents from a time series *Physica* D **16** 285–317 (Ch 7)

Wolfram S 1984 Universality and complexity in cellular automata *Physica* D **10** 1–35 (Ch 6)

——1986 Cellular automaton fluids 1: basic theory *J. Stat. Phys.* **45** 471–526 (Ch 6)

Wu R-S and Lai Y-S 1994 An observation on the fractal characteristics of individual river plan forms *Fractals Natural Appl. Sci.* **41** 441–51 (Ch 3)

Xie F and Hu G 1995 Periodic solutions and bifurcation behaviour in the parametrically damped two-well Duffing equation *Phys. Rev.* E **51** 2773–8 (Ch 6)

Xie H 1995 Effects of fractal crack *Theor. Appl. Fracture Mech.* **23** 235–44 (Ch 3)

Xu T, Moore I D and Gallant J C 1993 Fractals, fractal dimensions and landscapes—a review *Geomorphol.* **8** 245–62 (Ch 4)

Yagisawa K, Kambara T and Naito M 1994 Chaos in the model of repetitive phase transitions with hysterisis: application to the self-sustained potential oscillations of lipid-bilayer membranes induced by gel–liquid-crystal phase transitions *Phys. Rev.* E **49** 1320–35 (Ch 6)

Yamada T and Fujisaka H 1984 Coupled chaos *Chaos and Statistical Methods: Proc. 6th Kyoto Summer Inst. (Kyoto, 1983)* ed Y Kuramoto (Berlin: Springer) pp 167–9 (Ch 6)

Yang W, Ding E-J and Ding M 1996 Universal scaling law for the largest Lyapunov exponent in coupled map lattices *Phys. Rev. Lett.* **76** 1808–11 (Ch 7)

Yin Z-M 1996 New methods for simulation of fractional Brownian motion *J. Comput. Phys.* **127** 66–72 (Ch 4)

Young I M and Crawford J W 1991 The fractal structure of soil aggregates: its measurement and interpretation *J. Soil Sci.* **42** 187–92 (Ch 3)

Zaslavsky G M, Sagdeev R Z, Usikov D A and Chernikov A A 1991 *Weak Chaos and Quasi-Regular Patterns* (Cambridge: Cambridge University Press) (Ch 6)

Zeni A R and Gallas J A C 1995 Lyapunov exponents for a Duffing oscillator *Physica* D **89** 71–82 (Ch 7)

Zhao J Y and Hahn E J 1993 Subharmonic, quasi-periodic and chaotic motions of a rigid rotor supported by an eccentric squeeze film damper *Proc. Inst. Mech. Eng.* **207** 383–92 (Ch 6)

Zosimov V V and Naugol'nykh K A 1994 Fractal structure of large-scale variability of wind driven waves according to laser-scanning data *Chaos* **4** 21–4 (Ch 3)

Zumofen G and Klafter J 1994 Random walks in the standard map *Europhys. Lett.* **25** 565–70 (Ch 4)

Index